SpringerBriefs in Biology

More information about this series at http://www.springer.com/series/10121

Nobuyuki Inui

Systematic Changes in Body Image Following Formation of Phantom Limbs

Springer

Nobuyuki Inui
Laboratory of Human Motor Control
Naruto University of Education
Naruto-shi, Tokushima
Japan

ISSN 2192-2179   ISSN 2192-2187   (electronic)
SpringerBriefs in Biology
ISBN 978-981-10-1459-8   ISBN 978-981-10-1460-4   (eBook)
DOI 10.1007/978-981-10-1460-4

Library of Congress Control Number: 2016943184

Printed on acid-free paper

This Springer imprint is published by Springer Nature
The registered company is Springer Science+Business Media Singapore Pte Ltd.

Preface

About 10 years ago, I read the book *Touch,* written in Japanese by Dr. Yoshiaki Iwamura, who had found that receptive fields of neurons in somatic sensory cortices in monkeys are represented hierarchically. As reports of an interesting relationship between kinesthesia and body image, works of Drs. D.I. McCloskey and Simon C. Gandevia were cited in the book. I was inspired by these studies on body image. In particular, I was interested in Prof. Gandevia's work on the relationship between body image and kinesthesia. After I read their papers and reviews of kinesthesia, I went to the Prince of Wales Medical Research Institute (Neuroscience Research Australia at present) in Sydney in 2007. In the laboratory of Prof. Gandevia, I conducted experiments shown in Sects. (2.1), (2.2), and (2.6) of Chap. 2 with Drs. Lee Walsh and Janet Taylor. After coming back from Sydney, I extended a series of experiments shown in Sects. (2.3), (2.4), and (2.5) of Chap. 2 and in Chaps. 3 and 4 in my laboratory with a graduate student, Junya Masumoto (Ph.D. at present).

Perception of the body consists of information from multiple sensory modalities, and the perception forms a body image. We have to block sensory information in part and analyze processes forming a body image. Prof. Gandevia provided me a method to block sensory information in part. In addition, I was taught to examine psychological phenomena using physiological methods.

Here, I present the fruit of our studies on body image in my laboratory at Naruto University of Education following the study in Sydney.

Naruto, Japan   Nobuyuki Inui
October 2015

v

Contents

Chapter 1
Introduction

Abstract This monograph reviews our recent studies on body image. This chapter describes a brief history of studies on body image and phantom limbs and the aim of this monograph. Recently, "online" and "offline" representations of the body are proposed. The online representation is provided by inputs such as vision, touch, and proprioception, and it updates perception of the body from moment to moment. The offline representation is constructed, in part, from current sensory inputs and, in part, from stored memories. The concept of offline representation is useful to account for phenomena such as the development of phantom limbs after amputation. On the other hand, the attempt to produce an acute block of peripheral nerves to a limb to create an experimental phantom provides insight into how an online representation of the body might be constructed. Our studies thus use an ischemic block of a limb to study the mechanisms of changes in body image. A series of our studies shows that the body image is constantly being updated according to proprioceptive, tactile, and visual information.

Keywords Body image • Body schema • Phantom limbs • Ischemic anesthesia

Over 100 years, the pioneering concept of body schema (Head and Holmes 1911) has led to a proposal of two distinct body representations: body image and body schema (de Vignemont 2010; Proske and Gandevia 2012). The ability to feel and move a limb requires the brain to use an "image" of the body which constitutes part of our self-awareness (James 1890). Such a representation underlies our conscious experience, and it can become distorted in many neurological and psychiatric conditions (Frith et al. 2000). To make accurate movements, information is required about the sizes and lengths of body segments and the position and movement of joints (Sanes et al. 1984; Sainburg et al. 1993). This is derived from a modifiable body "schema" (Head and Holmes 1911; Cole and Paillard 1995; Giummarra et al. 2007). The terminology for these central representations is controversial (de Vignemont 2010; Longo et al. 2010).

The body schema derives from somatosensory, visual, and vestibular inputs and is generated in the parietal cortex as well as in more distributed cortical networks (Schwoebel and Coslett 2005). A function of the parietal cortex is proposed to update and integrate actual and predicted (efference copy) sensory feedback about

© The Author(s) 2016

N. Inui, *Systematic Changes in Body Image Following Formation of Phantom Limbs*,
SpringerBriefs in Biology, DOI 10.1007/978-981-10-1460-4_1

the sizes and lengths of body segments and the position and movement of joints (Shadmehr and Krakauer 2008). The efference copy reaches the premotor cortex to drive movement corrections when errors arise (Wolpert et al. 1995). Thus, changes in body schema could be occurring with the parietal cortex along with the premotor cortex.

Carruthers (2008) further proposes "online" and "offline" representations of the body. The online representation is provided by inputs such as vision, touch, and proprioception, and it updates perception of the body from moment to moment. The offline representation is constructed, in part, from current sensory inputs and, in part, from stored memories. The concept of offline representation is useful to account for phenomena such as the development of phantom limbs after amputation.

The phenomenon of phantom limbs has been already reported in the middle ages. European folklore glorified the restoration of sensation in amputated limbs in soldiers. In the sixteenth century, the French military surgeon Ambroise Pare noted many cases of phantom limbs among soldiers returning from European battlefields. As the modern clinical investigation of phantom limbs, the American neurologist Silas Weir Mitchell wrote up interviews with thousands of amputees after the civil war. He coined the term "phantom limb." More than 70 % of patients perceived their phantom limbs painful immediately after surgery. About 60 % of the patients felt the throbbing for years (Nicolelis 2011; Ramachandran and Blakeslee 1998). Wall (2000) early proposed that the phantom limb phenomenon originated as a result of spurious activity generated by severed nerve fibers in the scarred region of the sump. However, when many neurosurgeons cut the sensory nerves leading to the spinal cord, the phenomenon persisted. Neuroscientists thus rejected the notion that neuromas, or any other abnormality at the level of peripheral nerves, could explain the richness of symptoms of the phantom limb syndrome.

That picture started to change when Merzenich et al. (1983) found that traumatic amputation of the middle finger of an adult monkey led to a remarkable functional reorganization of the somatotopic map in the primary somatosensory cortex. In monkeys whose entire arm had been deafferented for many years, Pons et al. (1991) also found that long-term deafferentation prompted a widespread reorganization, in which the neurons previously assigned to the hand switched to react to singles from the face. These findings gave the American neuroscientist Ramachandran a useful suggestion for an explanation for the occurrence of phantom limbs in amputees. Then, Ramachandran et al. (1992) found that tactile stimulation of specific points on the face elicited sensations at specific points on the phantom hand. In addition, Flor et al. (1995) reported that there was a strong correlation between the amount of functional cortical restructuring and the magnitude of phantom pain employing magnetoencephalography.

Ramachandran and Hirstein (1998) have pointed out two potential explanations for the occurrence of phantom limbs in amputees. One explanation is the occurrence of plastic reorganization of the topographic body map present in the somatosensory cortex of patients whose limbs had been amputated. This explanation is thought of with two possibilities (Ramachandran and Blakeslee 1998). First, the reorganization could involve sprouting: the growth of new branches from nerve fibers that normally innervate the face area toward neurons in the hand area in the cortex. If this possibility was true, this would be quite remarkable because it is difficult to see how highly organized sprouting could take place over relatively long distances and in such a short period. Second, although there is a tremendous redundancy of connections in the normal adult brain, most of them are nonfunctional or have no obvious function. Like amputation, they may be called into action only when needed. Thus, there might be sensory inputs from the face to the brain's face map and to the hand map area as well. If so, we must assume that this hidden input is ordinarily inhibited by the sensory fibers arriving from the real hand. But when the hand is removed, this silent input originating from the skin on the face is unmasked and allowed to express itself so that touching the face now activates the hand area and leads to sensations in the phantom hand.

Another explanation for the occurrence of phantom limbs is associated to the function of the parietal cortex. If a person moves his limbs with closed eyes, he feels the position and movement of his limbs. The body image is stored in the parietal cortex with various sensory information. If a human performs movements, the motor command from the motor cortex reaches the parietal cortex (efference copy) as well as the spinal cord, producing the perception of limb movements. Thus, amputated limbs are perceived as moved by the body image in the parietal cortex. Based on the brain's internal workings, Ramachandran (1996) contrived mirror box therapy to practice calming a phantom arm. In the mirror box, a mirror is positioned vertically in the center of a box, the amputated limb and phantom are placed on one side out of view, and the intact limb is placed so that the amputee can see its reflection. During the therapy, the amputee observes the reflection of his normal limb moving in the mirror in order to induce the visual illusion that he has two intact limbs and that the intact limb is superimposed onto the felt position of his phantom limb. When patients moved their intact arms, they felt the phantom arm obey these same motor commands. Six of the patients who used the mirror box said they could also feel their phantom arm moving, generating the impression that both arms could now be moved. Four of the patients used this newfound ability to relax and open a clenched phantom hand, providing relief from painful spasms. In one patient, a routine of ten minutes of practice per day with the mirror caused his phantom arm and elbow to disappear completely within 3 weeks (Ramachandran et al. 1995).

On the other hand, the attempt to produce an acute block of peripheral nerves to a limb to create an experimental phantom provides insight into how an online representation of the body might be constructed. Phantom limbs generally adopt a habitual posture or a posture that resembles that prior to amputation (Ramachandran and Hirstein 1998; Paqueron et al. 2004). Melzack and Bromage (1973) ask prone participants to describe the position of a phantom arm after an anesthetic nerve block at the brachial plexus. The phantom arm lies at the side of the body, above the lower abdomen, or above the lower chest. Orderly perceived changes in posture occur spontaneously between these three positions during the block. However, the extent to which the phantom limbs depend on afferent input and the mechanisms underlying any perceptual changes and distortion of the phantom limbs are poorly understood. Slightly, Gentili et al. (2002) suggest that the posture of an experimental phantom is close to the posture of the limb when the nerve block is induced. It is anticipated that the initial posture of an experimental phantom has an influence on the final posture of the phantom. We thus determine that the initial position of a limb before anesthesia is critical for the subsequent illusory changes in posture in a series of our studies (Inui et al. 2011, 2012a, b; Inui and Masumoto 2013). In addition, we determine what a key parameter is for the illusory changes in limb position (for review, Inui and Masumoto 2014).

This monograph reviews our recent studies on body image. The monograph is divided into four chapters. In Chap. 1 the author describes a brief history of studies on body image and phantom limbs and the aim of this monograph. In Chap. 2 the author first reviews systematic changes in the perceived posture of an experimental phantom hand following cuff inflation of the upper arm (Inui et al. 2011). Second, the author reports changes in sensation following an ischemic nerve block (Inui et al. 2011). Third, the author points out that the extreme posture of a limb is essential for systematic changes in the perceived posture during the block (Inui et al. 2012b). Fourth, the author describes systematic changes in the perceived posture by interactions between biarticular muscles across two joints during the block (Inui et al. 2012a). Fifth, the author quantitatively describes effects of visual information on the perceived posture of the foot and leg at the end of the block (Inui and Masumoto 2013). Sixth, the author describes changes in the perceived size of the hand following an ischemic nerve block (Inui et al. 2011). Chapter 3 reports that, while the right hand and forearm constituted an incomplete phantom limb from 10 to 20 min after the block, the stick fixed to the hand was perceived as having moved (Inui and Masumoto 2015). Chapter 4 finds that, in a virtual environment, whereas the proprioceptive estimate of arm position was gradually adapted to match the visual estimate, the visual estimate was very gradually adapting to match the proprioceptive estimate (Masumoto and Inui 2015). As a conclusion, Chap. 5 proposes that the key parameter for the illusory changes in limb position may be the difference in discharge rates between afferents in flexor and extensor muscles at a joint.

References

Carruthers G (2008) Types of body representation and the sense of embodiment. Conscious Cogn 17:1302–1316

Cole J, Paillard J (1995) Living without touch and peripheral information about body position and movement: studies with deafferented participants. In: Bermudez JL, Marcel A, Eilan N (eds) The body and the self. MIT Press, Cambridge, MA, pp 245–266

de Vignemont F (2010) Body schema and body image—pros and cons. Neuropsychologia 48:669–680

Flor H, Elbert T, Knetch S, Wienbruch C, Pantev C, Birbaumer N, Larbig W, Taub E (1995) Phantom limb as a perceptual correlate of cortical reorganization following arm amputation. Nature 375:482–484

Frith CD, Blakemore SJ, Wolpert DM (2000) Abnormalities in the awareness and control of action. Philos Trans R Soc Lond B Biol Sci 355:1771–1788

Gentili ME, Verton C, Kinirons B, Bonnet F (2002) Clinical perception of phantom limb sensation in patients with brachial plexus block. Eur J Anaesth 19:105–108

Giummarra MJ, Gibson SJ, Georgiou–Karistianis N, Bradshaw JL (2007) Central mechanisms in phantom limb perception: the past, present and future. Brain Res Rev 54:219–232

Head H, Holmes G (1911) Sensory disturbances from cerebral lesions. Brain 34:102–254

Inui N, Masumoto J (2013) Effects of visual information on perceived posture of an experimental phantom foot. Exp Brain Res 226:487–494

Inui N, Masumoto J (2014) Systematic changes in body image following ischemic nerve block. Compr Psychol 3:11

Inui N, Masumoto J (2015) Perceptual changes of interaction between hand and object in an experimental phantom hand. J Mot Behav 47:81–88

Inui N, Walsh LD, Taylor JL, Gandevia SC (2011) Dynamic changes in the perceived posture of the hand during ischaemic anaesthesia of the arm. J Physiol 589:5775–5784

Inui N, Masumoto J, Beppu T, Shiokawa Y, Akitsu H (2012a) Loss of large-diameter nerve sensory input changes the perceived posture. Exp Brain Res 221:369–375

Inui N, Masumoto J, Ueda Y, Ide K (2012b) Systematic changes in the perceived posture of the wrist and elbow during formation of a phantom hand and arm. Exp Brain Res 218:487–494

James W (1890) Principles of psychology. Henry Holt, New York

Longo MR, Azanon E, Haggard P (2010) More than skin deep: body representation beyond primary somatosensory cortex. Neuropsychologia 48:655–668

Masumoto J, Inui N (2015) Visual and proprioceptive adaptation of arm position in a virtual environment. J Mot Behav 47:483–489

Melzack R, Bromage PR (1973) Experimental phantom limbs. Exp Neurol 39:261–269

Merzenich MM, Kass JH, Wall RJ, Nelson RI, Sur M, Felleman D (1983) Topographic reorganization of somatosensory cortical areas 3B and 1 in adult monkeys following restricted deafferentation. Neuroscience 8:33–55

Nicolelis M (2011) Beyond boundaries: the new neuroscience of connecting brains with machines — and how it will change our lives. St. Martin's Griffin, New York, pp 56–57

Paqueron X, Leguen M, Gentili ME, Riou B, Coriat P, Willer JC (2004) Influence of sensory and proprioceptive impairment on the development of phantom limb syndrome during regional anesthesia. Anesthesiology 100:979–986

Pons T, Preston E, Garraghty AK (1991) Massive cortical reorganization after sensory deafferentation in adult macaques. Science 252:1857–1860

Proske U, Gandevia SC (2012) The proprioceptive senses: their roles in signaling body shape, body position and movement, and muscle force. Physiol Rev 92:1651–1697

Ramachandran VS (1996) Phantom limbs, neglect syndromes, repressed memory and Freudian psychology. In Sporns O, Tonini G (eds) Selectionism and the brain. Int Rev Neurobiol 37:291–333

Ramachandran VS, Blakeslee S (1998) Phantoms in the brain: probing the mysteries of the human mind. William Morrow, New York

Ramachandran VS, Hirstein W (1998) The perception of phantom limbs. Brain 121:1603–1630

Ramachandran VS, Rogers–Ramachandran D, Stewart M (1992) Perceptual correlates of massive cortical reorganization. Science 258:1159–1160

Ramachandran VS, Rogers–Ramachandran D, Cobb S (1995) Touching the phantom limb. Nature 377:489–490

Sainburg RL, Poizner H, Ghez C (1993) Loss of proprioception produces deficits in interjoint coordination. J Neurophysiol 70:2136–2147

Sanes JN, Mauritz KH, Evarts EV, Dalakas MC, Chu A (1984) Motor deficits in patients with large-fiber sensory neuropathy. Proc Natl Acad Sci U S A 81:979–982

Schwoebel J, Coslett HB (2005) Evidence for multiple, distinct representations of the human body. J Cogn Neurosci 17:543–553

Shadmehr R, Krakauer JW (2008) A computational neuroanatomy for motor control. Exp Brain
   Res 185:359–381
Wall PD (2000) Pain: the scientific of suffering. Columbia University Press, New York
Wolpert DM, Ghahramani Z, Jordan MI (1995) An internal model for sensorimotor integration.
   Science 269:1880–1882

Chapter 2
Formation of Phantom Limbs Following Ischemic Nerve Block

Abstract The brain needs body image to plan movement. We use ischemic anesthesia of a limb to study the mechanisms of changes in body image (Sects. 2.1, 2.3, and 2.5). First, if the fingers, wrist, elbow, knee, and ankle are extended before and during anesthesia, the perceived phantom limbs are flexed at the joints and vice versa. However, when the hand is held in the mid-position before and during the anesthesia, the position of the wrist is perceived to be in the same position (Sect. 2.3). Hence, the fully flexed or extended position of a limb was essential for systematic changes in the perceived posture of the limb during the anesthesia. In addition, if the actual wrist was fully extended while the actual elbow was fully flexed, then the perceived position of the wrist moved toward flexion and that of the elbow moved toward extension and vice versa (Sect. 2.4). Following the loss of the afferent signal coming from the main muscles acting at the two joints, the two perceived postures moved toward the opposite direction independently. Second, perceived hand size increases by $34 \pm 4\%$ (mean $\pm 95\%$ confidence interval) as anesthesia develops (Sect. 2.6). Third, the start of these perceptual changes occurs when input from large-diameter sensory nerve fibers is declining (Sect. 2.2). Fourth, at the end of the ischemic block, when participants are allowed to see their foot, its perceived position reverts to that indicated by them earlier (Sect. 2.5).

Keywords Phantom limb • Ischemic anesthesia • Proprioception • Tactile threshold • Biarticular muscle • Perceived size • Vision

2.1 Development of a Phantom Hand

Even when the hand is stationary we know its position. This information is needed by the brain to plan movements. If the sensory input from a limb is removed through an accident or an experiment with local anesthesia, then a "phantom" limb commonly develops. A central body representation provides a template for perception of phantom limbs after amputation and deafferentation. Acute ischemic anesthesia of the upper limb was used to generate a phantom hand (Gandevia et al. 2006; Walsh et al. 2010), and we studied the perceptual changes as it developed. Our aim for this section was to examine how an online representation of the body might

© The Author(s) 2016

N. Inui, *Systematic Changes in Body Image Following Formation of Phantom Limbs*, SpringerBriefs in Biology, DOI 10.1007/978-981-10-1460-4_2

Fig. 2.1 Method used to
produce ischemic anesthesia
and paralysis of the right
forearm and hand using a
pressure cuff on the upper
arm. To test position sense,
the right wrist and hand can
be placed in specific postures
by an experimenter, and
participants use their left
hand to move the fingers of a
wooden hand and pointer on
the model to indicate its
perceived position

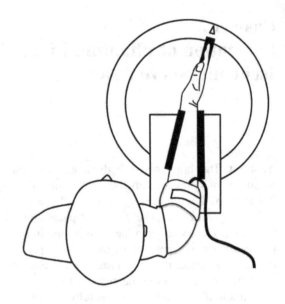

be constructed and determine whether the initial posture of the wrist and fingers
before anesthesia systematically changes the final posture of the phantom hand.

This form of anesthesia was chosen because it develops slowly with loss of
conduction in large-diameter axons occurring before that in small-diameter axons.
The various classes of fibers in peripheral nerves differ in fiber diameter and speed
of conduction. In addition, the various classes of peripheral nerves differ in their
sensitivity to hypoxia and anesthesia. Local anesthesia depresses transmission in
the group C fibers before they affect group A touch fibers. In contrast, pressure on
a nerve can cause loss of conduction in large-diameter motor, touch, and pressure
fibers, while pain sensation remains relatively intact.

Experiments were performed on ten adult male and female participants who were
healthy without any apparent neurological disorders (age range, 24–54 years). The
participants were comfortably seated and the right hand held in mid-pronation in
a manipulandum that rotated in the flexion–extension plane of the wrist (Fig. 2.1).
Using the apparatus, the hand and arm up to above the elbow of the participants
were screened. This experiment was conducted twice in all ten participants, once
with the fingers taped fully extended to a plate attached to the manipulandum in full
extension and once with the fingers taped in full flexion and the wrist comfortably
flexed. Five participants were studied first in full extension, whereas the other five
were studied first in full flexion. The participants knew that their hands and arms
were fixed before and during anesthesia. However, because the participants put on
an eye mask before their hands and arms were fixed, they did not look at the initial
position of the limb.

A wide cuff with two chambers (Zimmer, Dover, OH, USA) was positioned on
the upper arm and connected to a regulated pressure source (Ulco Engineering,
NSW, Australia) so that the cuffs could be simultaneously inflated to 300 mmHg

in less than 1 s. This produces more consistent blocks than gradual manual inflation of a sphygmomanometer cuff.

To measure the development of the phantom hand, participants indicated the perceived angular positions of their phantom fingers and wrist using a wooden hand, the posture of which allowed positioning in flexion and extension at the metacarpophalangeal (MCP) and interphalangeal joints of the fingers. To assess perceived position at the wrist, a rotating pointer on the model was used. The model was positioned on the table about 20 cm in front of the participants and 30 cm to the left of the midline. Participants were asked to "signal the perceived finger positions of your right hand by using the wooden hand" and to "think about each finger individually." The wooden hand and wrist pointer were photographed with a digital camera mounted above the model to provide an indication of the perceived angular positions of the fingers and wrist. After each estimate of wrist and finger posture, the joints of the model hand were alternately placed in full extension or full flexion. Six points on the wooden hand were marked to measure joint angles using a software package (GNU Image Manipulation Program): the tip of the index, the distal interphalangeal (DIP) joint, the proximal interphalangeal (PIP) joint, the metacarpophalangeal (MCP) joint, the metacarpal base, and the wrist. Data for the angles of the index finger were analyzed quantitatively. However, similar changes in angles occurred for the other fingers. Wrist angle was deemed to be at 180° with the metacarpal in line with the forearm. Angles <180° indicate flexion and >180° extension. Full extension of the fingers was set at 180°.

Before inflation of the cuff, no change of the perceived position was confirmed when the wrist and fingers were maintained in the test postures. The control trials were carried out four times on the participants with their eyes closed for 10 min. The control measurements of touch and posture were conducted at the start of the control trial and intervals of 2 min, and shortly thereafter the measurements of touch and posture were conducted at the start of inflation and intervals of 5 min. Voluntary movement of the wrist and elbow abolished at the end of the block (at 35–40 min). Because attempted voluntary movements during paralysis may generate perceptions of movement in the direction of the effort (Gandevia et al. 2006; Walsh et al. 2010), in this study we confirmed that paralysis was complete only at the end of the block. In addition, a control study in four participants revealed that there was no significant adaptation of the participant's perceived position of their intact wrist and fingers over a period of 40 min when the hand was maintained in the test posture with the wrist and fingers held comfortably extended.

The first study indicated that the initial posture of the hand (held for 10 min prior to cuff inflation) at either full flexion or full extension was critical for the change in perceived joint angles of a phantom hand. Hence, in two participants, we repeated the block with the hand in the fully flexed position for 10 min and then moved to the fully extended position only at the time at which the cuff was inflated. Again, participants signaled the perceived angular positions of their phantom, and simple sensory testing for light touch was conducted after cuff inflation.

In the main experiment, the hand and wrist were held in a fully flexed or fully extended position for 10 min prior to cuff inflation. Pilot testing revealed

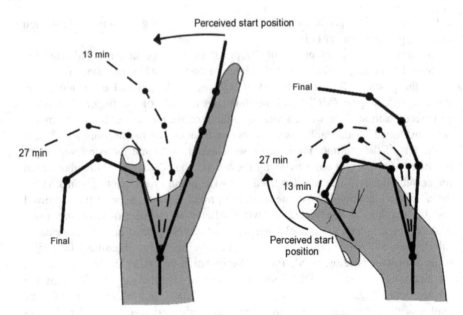

Fig. 2.2 Actual and perceived posture of the hand during an ischemic block. Data are obtained from ten participants for the two versions of the main experiment (wrist and hand extended on *left*, wrist and hand flexed on *right*). *Shaded* hands show the true maintained posture of the hand. *Thick black lines* show the perceived starting position of the hand prior to cuff inflation and the perceived final position of the phantom hand at the end of the block. The *dashed lines* are the mean perceived position of the developing phantom at intermediate times of 13 and 27 min. Data at these times were determined from the regression fitted for each participant. Time zero is the time of cuff inflation. The direction of the perceived change is given by the *curved arrow* (Inui et al. 2011, Redrawn with permission from the Journal of Physiology)

that over this time the participants correctly signaled that their fingers were flexed or extended with minimal adaptation. However, the major new finding was that, after cuff inflation, despite the position of the hand and wrist being fixed (in either flexion or extension), the perceived position of the wrist and fingers changed systematically. This change was evident after 10–15 min, and surprisingly, the direction of the change depended on the posture in which the hand was held. If the actual hand posture was one of extension, then the perceived position of the wrist and fingers moved toward flexion. Conversely, if the posture was one of flexion, then the perceived position of the wrist and fingers moved toward extension. This slow change in position did not involve hand postures which were anatomically impossible, and changes were in the same direction, but not always of the same magnitude for each of the fingers. In all participants, the perceived position moved away from the real position of the hand. Figure 2.2 illustrates changes in the perceived posture of the hand during the block. In Fig. 2.3, mean data (±95 % confidence intervals) are shown for the absolute changes in perceived joint angle (at the wrist and three index finger joints). They emphasize that the final perceived

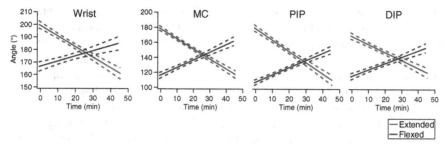

Fig. 2.3 Perceived angle of the wrist and index finger joints during an ischemic block. The *solid lines* represent regression lines of the average of all ten participants with 95 % confidence intervals (*dashed lines*) for the wrist, metacarpophalangeal joint (MC), proximal interphalangeal joint (PIP), and distal interphalangeal joint (DIP). 180° represents a straight joint (Modified from Inui et al. 2011)

postures differed depending on the maintained hand posture. Although hand position was maintained throughout the period of cuff inflation, the final perceived position of the hand was more flexed when the hand was held in extension than when it was held in flexion.

To determine whether the changes in posture of the hand once the cuff was inflated depend on it being held in one posture for 10 min prior to inflation, we repeated the block in two participants. Again, the hand was fixed in the fully flexed position for 10 min, but just before the cuff was inflated, the hand was moved into the fully extended position and held there. The perceived posture of the wrist and fingers curled into flexion in a qualitatively similar way to when the hand had been held for 10 min in the extended posture prior to cuff inflation. Hence, holding the wrist and hand positions constant for 10 min prior to cuff inflation is not critical for generation of the subsequent illusory changes in position.

What is the mechanism for the progressive change in posture as the ischemic block develops? The brain has access to different central representations of the limb (Sirigu et al. 1991; Aglioti et al. 1995; Kammers et al. 2006; Longo et al. 2010) which subserve different sensory, motor, and other functions, but those with sensorimotor roles depend on the background sensory input. Psychophysical studies have shown that the sensory input generating the proprioceptive image of the body is likely to be derived from specialized muscle receptors, particularly muscle spindle endings (Goodwin et al. 1972; Roll and Vedel 1982; Gandevia 1985) with a contribution also from cutaneous stretch receptors (Edin and Johansson 1995; Collins et al. 2005). This image is sustained over the minutes prior to cuff inflation and induction of anesthesia. As the input from sensory receptors "fades" during the anesthesia, the perceived hand moves away from its initial maintained position (cf. Paqueron et al. 2004b).

Then, there is a peripheral contribution to the illusory changes in wrist and finger posture. This would require a differential effect of cuff inflation on the population response of activity in afferents signaling joint position. When the hand is fully flexed, it is likely that the firing rate of receptors which encode flexion will be greater than that of receptors signaling extension. For example, muscle spindles in the extensor muscles of the wrist and fingers will fire because they are stretched, whereas those in the flexors will fire at low rate or be silent. Similarly, proprioceptive cutaneous afferents innervating skin stretched over the dorsum of the hand will fire more than those on the palmar surface (Edin 1992). As conduction block progressively develops, the reduction in the firing rate for the population of "flexor encoding" receptors will be greater than for the extension encoding receptors. Such a differential response would lead to a perception that the hand is moving away from its initial extreme position. We propose that the progressive loss of sensory input which signals that the fingers are extended is interpreted as the phantom hand folding into flexion and vice versa.

This process may coexist with other short-term central changes accompanying the ischemic block (e.g., Brasil-Neto et al. 1993). Loss of sensory input caused by an ischemic block could lead to the release of afferent inhibition which might produce the increase in cortical excitability observed after an ischemic block (Vallence et al. 2012). The release of afferent inhibition is supported by a decrease in GABA levels in the primary motor cortex measured using magnetic resonance spectroscopy during an ischemic block (Levy et al. 2002). These previous studies thus suggest that the change in body image from the extreme posture during the ischemic block is induced by changes in cortical excitability.

Additional mechanisms are needed to explain the fact that the final postures produced by holding the hand in extension or in flexion are not the same but have actually "crossed over" (see Fig. 2.3). For example, the changing peripheral inputs during the ischemic block from a strong flexion population input or a strong extension population input may induce different levels and degrees of adaptation in channels signaling flexion and extension. For some aspects of proprioception, there is evidence for the operation of directionally sensitive opponent channels (Seizova-Cajic et al. 2007).

2.2   Changes in Sensation Following Ischemic Anesthesia

In this section the author compares the timing of impairment of different sensory modalities during the development of the ischemic block. During the block there is a progressive degradation in touch sensation. Thus, in Sect. 2.1, the changes in body image appear to occur as input from large-diameter sensory nerve fibers is declining.

The experiment was conducted by the same ten participants as Sect. 2.1. Three sites for sensory testing were marked: the dorsum of the thumb, just proximal to the dorsum of the wrist, and the volar surface of the forearm just distal to the elbow. Sensory tests included assessments of tactile sensation with von Frey hairs and the ability to detect light touch with a cotton swab. To assess the ability to detect a light touch, the three testing sites were touched by a cotton swab in random order, and the participants were asked to identify the site of the touch. Before and after inflation of the cuff, the measurement of touch was conducted in the same procedure as Sect. 2.1.

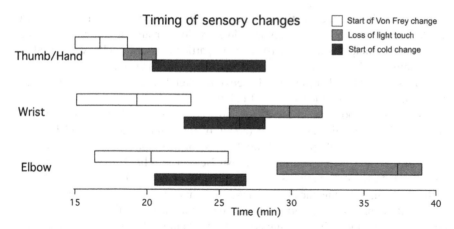

Fig. 2.4 Timing of changes in sensation following an ischemic block of the upper arm. Data are shown for assessments made over the base of the thumb (*upper panel*), just proximal to the wrist (*middle panel*), and just proximal to the elbow (*lower panel*). Mean and interquartile range plotted for the start of a change in tactile threshold (von Frey), the loss of light touch, and the onset of elevation in cold thresholds (Inui et al. 2011, Redrawn with permission from the Journal of Physiology)

In addition, temperature sensation was assessed. Again, the tests were conducted four times before inflation and at 5-min intervals thereafter. A computer-driven thermode (TSA-2001 Model TSA II, Medoc Ltd., Ramat Yishai, Israel) was used to assess thermal and pain thresholds. A 30 × 30-mm thermode was placed on three sites: on the hand just proximal to the thumb, just proximal to the wrist, and just distal to the elbow. A cold detection threshold and the threshold for heat pain were assessed. Participants were given thermal stimulation (from a baseline of 30 °C, rise and return rates of 1.0 °C/s (cooling) and 1.5 °C/s (heat pain)) and were asked to say "stop" as soon as cool or heat pain was detected. The same wrist and elbow locations were used for thermal testing as for sensory testing. However, due to the size of the thermode, the most distal site was not on the dorsum of the proximal phalanx of the thumb, but about 2 cm proximal and lateral to the first metacarpophalangeal joint.

Measurements of touch in the hand, just proximal to the wrist, and just distal to the elbow change systematically. Cutaneous touch sensation is impaired progressively, beginning distally (Fig. 2.4). The initial impairment (an increase in tactile threshold) begins at 16.5 [15.0–18.5] (median [interquartile range]) min for the hand and at 20.5 [16.5–25.5] min for the elbow. Complete loss of detection of light touch follows the same pattern, again occurring earlier at the hand (19.5 [18.3–20.6] min) than at the elbow (37.5 [29–39] min). About 30 min after inflation, voluntary movement was abolished for wrist and finger movements. At 20–25 min, thresholds for cold sensation begin to change throughout the arm while we use a computer-driven thermode to assess thermal pain thresholds. Cold is initially reported at a mean of 28.3°–28.8° at the three sites. These temperatures drop by 5°–9° at the end of cuff inflation, with the change being slightly greater distally than proximally. Then, cold stimuli are commonly reported as painful. Changes in tactile thresholds precede changes in cold sensation for all ten participants at the hand, nine participants at the wrist, and six participants at the elbow. Even after 30–40 min of ischemia, high skin temperatures still elicit a distinct sensation of heat pain. This index of C fiber function changes only slightly at each site. The thresholds rise from 43° to 44° by the end of cuff inflation.

The perceived postures of the finger and wrist are changed from 10 to 15 min after inflation. For this short period (about 10 min), the right hand and forearm constitute an incomplete phantom limb because of incomplete loss of detection of touch in the hand, wrist, and elbow. Brasil-Neto et al. (1993) have reported that corticospinal excitability increases in muscles proximal to an ischemic nerve block within 7–8 min following the block. The starting time of the increase in corticospinal excitability approximately corresponds to that of a change in the finger and wrist postures in this study. The existing cortical circuits might be disinhibited from this time after the block. This new finding that the incomplete phantom hand is perceived as having moved suggests a change in body image on the basis of the rapid changes of the cortical circuits.

In the conventional view (Gasser and Erlanger 1929; Mackenzie et al. 1975), pressure on a nerve can cause loss of conduction in large-diameter motor, touch, and pressure fibers, while pain sensation remains relatively intact. Conversely, local anesthetics depress transmission in the group C fibers before they affect group A touch fibers. Inui et al. (2011) assess thermal and pain thresholds, showing resistance to prolonged ischemia of small fibers (presumably group C fibers), which signal heat pain. Smaller fibers (presumably group Aδ fibers) also include the main population of cold (Johnson et al. 1973; Mackenzie et al. 1975), and by the end of the ischemic period (40 min), we find a marked impairment of cold thresholds (by 5°–10°). These observations make it likely that a change in body image can be initiated and driven by incomplete loss of background input from large-diameter sensory axons.

2.3   Extreme Postures Are Essential for Movements of a Phantom

Inui et al. (2011) tracked changes in the perceived posture of the hand during ischemic anesthesia of the arm to determine whether the initial posture of the wrist and fingers before anesthesia systematically influenced the perceived posture of the hand. We found that if the wrist and fingers were extended fully before and during anesthesia, the final hand was perceived as bent at the wrist and fingers and vice versa. The original aspects of this study are the finding that, when the phantom forms, the final perceived position of the hand depends on its initial position and that there is a seamless perceived position for the hand.

In our previous study (Inui et al. 2011), the firing of afferents signaling a fully flexed or extended hand appeared to be high and the other very low so that

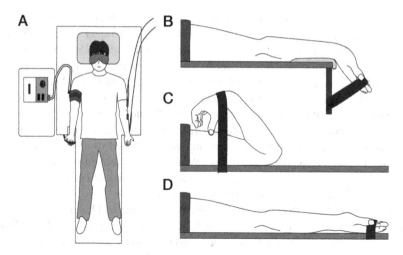

Fig. 2.5 Experimental setup. (**a**) Schematic of experimental setup. Participants comfortably lay on their backs on a table, and their right hands or arms were fixed by a tape with Velcro. (**b**) Full extension at the elbow and wrist joints. (**c**) Full flexion at the elbow and wrist joints. (**d**) Mid-position at the wrist and full extension at the elbow. Although the arm and hand were held straight by a wooden board, the arm was fully extended while the hand was not (Inui et al. 2012b)

disappearance of afferents mainly altered the high signal as the ischemic anesthesia progressed. However, it is predicted that if the hand is in mid-position, the firing of afferents signaling flexion and extension is balanced so that as afferent firing disappears it does so in a balanced way. Therefore, by holding the wrist in the fully extended, fully flexed, and mid-positions before and during the anesthesia, the study in this section demonstrates that a fully flexed or extended position of the wrist is essential for systematic perceived changes in the posture of the wrist when anesthesia sets in.

This study was performed on 20 adult male participants who were healthy and without any apparent neurological disorders (age range, 20–25 years). Ten participants participated in the first study, and the remaining ten participated in the second study. Participants lay comfortably on their backs on a table with their eyes closed and their right hands or arms fixed by Velcro (Magic Tape, Kuraray Co., Ltd., Tokyo, Japan, see Fig. 2.5a). The first study was conducted twice in all ten participants, once with the arm and hand fully extended (Fig. 2.5b) and once with the arm and wrist fully flexed (Fig. 2.5c). Five participants were studied first in full extension of the arm and hand, whereas the other five were studied first in full flexion of those. The second study was conducted once in all ten participants to examine what happened if the hand was held in mid-position before and during the anesthesia. The arm and hand were held straight and fixed to a wooden board on a table (Fig. 2.5d); while the arm was fully extended, the hand was held in the mid-position.

A wide cuff (width, 107 mm; Zimmer, Dover, OH, USA) was positioned on the right upper arm and connected to the automatic tourniquet system (Zimmer ATS 750) so that the cuff could be simultaneously inflated to 250 mmHg in less than 1 s. In the same procedure as Sect. 2.2 of this chapter, sensory tests included assessments of tactile sensation with von Frey hairs (North Coast Medical, Inc., Morgan Hill, CA, USA) and the ability to detect a light touch with a cotton swab. To measure perceived arm and hand postures during the development of the ischemic block, the participants were instructed to demonstrate the perceived positions of the right elbow and wrist using the left arm and hand attached to a flexible two-axis goniometer system (DKH, Tokyo, Japan). While the locations of the left and right hands and arms were symmetric with respect to the median line before each testing trial (see Fig. 2.5a), the left arm and wrist were extended.

The major finding is that, despite the position of the arm and hand being fixed (in either flexion or extension), after cuff inflation the perceived position of the elbow and wrist changed systematically in all participants. If the actual arm and hand were fully extended, then the perceived position of the elbow and wrist moved toward flexion. Conversely, if they were fully flexed, then the perceived position of the joints moved toward extension. As a new finding of this study, however, when the hand was held in the mid-position before and during the block, the position of the wrist was perceived to be in the same position. Figure 2.6 shows schematics of data (A) of perceived change in joint angle when the actual arm and hand was fully extended, data (B) of the perceived change in joint angle when the actual arm and hand was fully flexed, and data (C) of the perceived change in joint angle when the hand was held in the mid-position while the arm was fully extended. These systematic changes in position did not involve arm and hand postures that were anatomically impossible, and the changes were in the same direction. There was a seamless perceived position for the arm and hand. In addition, the final position of the phantom limb depended on its initial position.

Changes in tactile threshold (von Frey) following cuff inflation (Fig. 2.7) were similar to the changes in Sect. 2.2 of this chapter (Fig. 2.4). At the end of ischemia, while cutaneous touch sensation of the thumb and wrist was lost in all 20 cases of 10 participants, that of the elbow was not lost in 7 of 20 cases. Although the participants reported subjectively that some paresthesia began within 5 min for the hand and 10 min for the elbow after cuff inflation, the initial impairment (an increase in tactile threshold) began at 18.0 [15.0–20.0] (median [IQR]) min for the hand and at 27.5 [20.0–30.0] min for the elbow. Complete loss of detection of light touch followed the same pattern, again occurring earlier at the hand (21.0 [20.0–25.0] min) than at the elbow (34.5 [30.0–39.0] min).

A new finding of the present study is that the perceived position of the wrist was maintained in the mid-position when the hand was held in the same position before and during the anesthesia. In contrast, a fully flexed or extended hand became perceived as an extended or flexed phantom hand as ischemic anesthesia progressed, congruent with the previous study (Inui et al. 2011). The important difference between an extreme and mid-position is that in mid-position the firing of afferents signaling flexion and extension is balanced so that as afferent firing

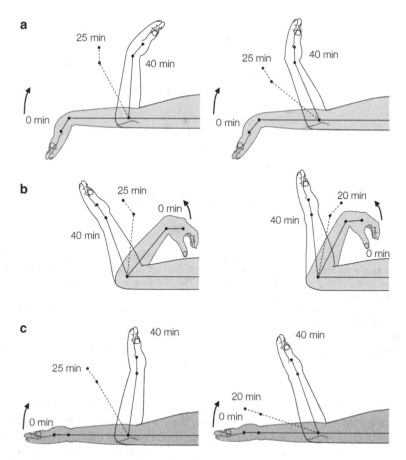

Fig. 2.6 Perceived position of a developing phantom during the ischemic block in both the first and second studies. (**a**) Data from one participant on *left* and data from ten participants on *right* for elbow and wrist extended in the first study. *Shaded* arms and hands show the true maintained posture of the arm and hand. *Solid lines* show the perceived starting position of the arm prior to cuff inflation and the perceived final position of the phantom arm at the end of the block [individual data at the elbow (180° to 86°) and wrist (212° to 138°), mean data at the elbow (184° to 124°) and wrist (210° to 158°)]. The *dashed lines* are the perceived position of the developing phantom at an intermediate time. Times are given relative to the time of cuff inflation. The direction of the perceived change is given by the *curved arrow*. (**b**) Data from one participant on *left* and data from ten participants on *right* for elbow and wrist flexed in the first study [individual data at the elbow (48° to 93°) and wrist (140° to 194°), mean data at the elbow (66° to 96°) and wrist (124° to 171°)]. (**c**) Data from one participant on *left* and data from ten participants on *right* for elbow and wrist extended in the second study [individual data at the elbow (181° to 82°) and wrist (180° to 188°), mean data at the elbow (180° to 110°) and wrist (179° to 183°)] (Inui et al. 2012b)

disappears it does so in a balanced way. In contrast, at an extreme position one signal will be high and the other very low so that disappearance of afferents will mainly alter the high signal. The perceived hand posture is determined largely by activity coming from afferents in muscle and skin stretched by extreme posture. As ischemic

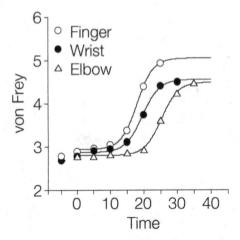

Fig. 2.7 Changes in tactile threshold (von Frey) over three sites for sensory testing in the first study. Data at each time are means of ten participants for both extension and flexion. Data on left are means of the control measurement for each site. The data were fitted to a sigmoid curve. While tactile threshold increased progressively as time passed, an increase in the threshold began distally. Because the sensory loss began distally during ischemia, mean data were lost distally. While the sizes of von Frey filament 1.65–6.65 correspond to 0.008–300 g of actual force, the set of filaments provides an approximately logarithmic scale of the force and a linear scale of perceived intensity. For example, a participant responded the 2.44 filament in the finger, the 2.36 one in the wrist, and the 2.36 one in the elbow during the control trials (Inui et al. 2012b)

anesthesia progresses, the stretch receptor activity falls, and this fall is interpreted by participants as a change in hand posture. However, the firing rate of the muscle spindles in both extensor and flexor when the hand is held in the mid-position may be less than that in the extreme position before and during the anesthesia. Hence, these results suggested that the loss of signals from muscle spindles in the fully flexed or extended position of a limb is essential for systematic changes in the perceived posture of a limb during an anesthesia.

As another new finding of the present study, the perceived position of the elbow changed systematically even when the somatosensory inputs from the arm were partially blocked. There were progressive changes in the perceived posture of the elbow and wrist different from the maintained posture: when the joints were flexed, they seemed to become extended, but when the joints were extended, they seemed to become flexed. While there was no significant difference between the elbow and wrist joints for the magnitude of the perceived changes in full extension, the magnitude of the perceived changes at the elbow reached about 80°. In full flexion, however, the magnitude at the elbow was less than that at the wrist, indicating that the peripheral nerves at the elbow were less blocked than those at the wrist at the end of the block. These results corroborated and extended the findings of our previous study (Inui et al. 2011) that there is no default position for the perceived hand and that the final position of the perceived hand depends on its initial position. Although Gentili et al. (2002) have reported that the posture of an experimental phantom is close to the posture of the limb at the start of the nerve block, their result was inconsistent with our result in the final position of the perceived hand. In addition, the magnitude of the perceived changes in the elbow was larger when they were extended than when they were flexed. This result was also consistent with the perceived changes in the wrist in our previous study (Inui et al. 2011) and indicated the operation of directionally sensitive opponent channels for some aspects of proprioception (Seizova-Cajic et al. 2007). Ribot-Ciscar and Roll (1998) and Bergenheim et al. (2000) have demonstrated that perception of movement is determined by the weighted input from different channels encoding direction. The changing peripheral inputs during the ischemic block from a strong flexion population input or a strong extension population input may induce different levels and degrees of adaptation in channels signaling flexion and extension.

Presumably, the progressive change in perceived posture underlies the short-term cortical reorganization following the loss of the large-diameter sensory input. Transient ischemic limb deafferentation appears to generate reversible short-term plasticity. When tested using transcortical magnetic stimulation during the course of ischemic nerve block, some studies observed a rapid increase in the motor cortical output to muscles proximal to the nerve block (Brasil-Neto et al. 1993; Ridding and Rothwell 1995) and an increase in cortical excitability obtained by repetitive cortical stimulation (Pascual-Leone et al. 1994; Ziemann et al. 1998). In a chronic deafferentation such as an amputation, Flor et al. (1995) found a strong positive correlation between the amount of cortical alteration and the magnitude of pain experienced after arm amputation using noninvasive neuromagnetic imaging techniques. This result indicated that phantom limb pain may be a consequence of plastic changes in the primary somatosensory cortex. Furthermore, using transcortical magnetic stimulation, Mercier et al. (2006) found sensations of movement in the phantom hand when the stimulation was applied over the presumed hand area of the motor cortex, suggesting that hand movement representations survive in the reorganized motor area of amputees. Such plastic changes may be enhanced by the selective loss of C fibers that occurs after amputation (Flor et al. 2006) because C fibers seem to have a special role in the maintenance of cortical maps (Calford and Tweedale 1991). In the present study, however, the 40-min ischemic block did not produce the complete loss of C fiber inputs but did produce the complete loss of large-diameter myelinated nerve inputs (also see Inui et al. 2011). In addition, because cutaneous touch sensation of the elbow was not lost in 7 of 20 cases, marked changes in the posture of the elbow were perceived by the partial loss of large-diameter sensory input. The results of present study thus suggested that the complete or partial loss of the large-diameter sensory input generated short-term cortical reorganization of the cortical map. By contrast, long-term reorganization of the cortical map generated by an amputation appears to depend on the loss of C fibers.

From these results, when should the term "phantom" be used? It is commonly applied to circumstances when there is a massive loss of afferent input, for example, from an amputation, a surgical nerve block, or a central lesion such as spinal injury. In many of these situations, some input from the periphery is preserved (Henderson and Smyth 1948; Hunter et al. 2003), and voluntary commands can still be issued to move and change the position of the paralyzed limb. In the present study in which there was a deliberately protracted loss of sensory inputs, we took an "experimental" phantom to mean a body part that was perceived when the large-diameter sensory input from that body part was completely gone. In the present study the initial perceived movements of the wrist from the real position were based on the real sensory changes. After 20–25 min of ischemia, the large-diameter sensory signals for the hand had completely stopped although a sensation of heat pain was still present even at the end of the block (Inui et al. 2011), and then the perception was an experimental phantom hand. Similarly, a change in the perceived elbow joint was based on real sensory changes because the large-diameter sensory signals from the arm had not completely stopped over the block. Nonetheless, the considerable change in the perceived elbow joint was highlighted at the end of the block. In particular, the magnitude of the perceived changes at the elbow in full extension was similar to that at the wrist, reaching about 80°.

2.4 Changes in Perceived Postures of a Phantom Forearm with Biarticular Muscles

In our previous studies (Inui et al. 2011, 2012b), the elbow, wrist, and fingers have been fixed in the same direction during an ischemic block. To further examine posture illusions during the block, we have to consider that many muscles are biarticular. For example, the flexors of both upper arm and forearm overlap across the elbow joint. When an ischemic block is performed with the wrist in full flexion, but with the elbow in full extension, the following alternative predictions could be made: either that there would be changes in the perceived posture due to the afferent signal coming from the main muscles acting at the two joints or that there would be no change in the perceived posture due to interactions between biarticular muscles across the two joints. Hence, this section examines what happens if the wrist is fixed in full extension while the elbow in full flexion before and during the anesthesia, and vice versa.

This study was performed on ten adult male participants who were healthy and without any apparent neurological disorders (age range, 20–25 years). The participants lay comfortably on their back on a table with their eyes closed and their right hand and arm fixed by two triangle bandages. The study was conducted twice in all ten participants, once with the wrist fully flexed and elbow fully extended and once with the wrist fully extended and elbow fully flexed (Fig. 2.8a). Five participants were studied first in full flexion of the wrist and full extension of the elbow, whereas the other five were studied first in full extension of the wrist and full flexion of the elbow.

In this section, the study was conducted by the same experimental procedure and measurement as Sect. 2.1 of this chapter. While a wide cuff (Zimmer, Dover, OH, USA) was positioned on the right upper arm and connected to the automatic tourniquet system (Zimmer ATS 750), the perceived positions of the right elbow and

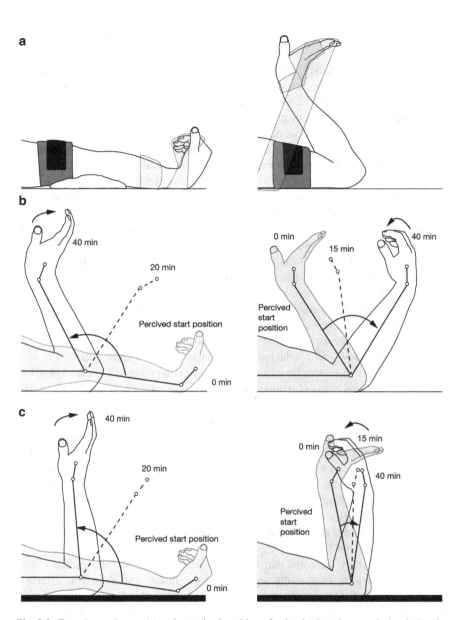

Fig. 2.8 Experimental setup (**a**) and perceived position of a developing phantom during ischemia (**b, c**). (**a**) Full flexion at the wrist and full extension at the elbow on *left* and full extension at the wrist and full flexion at the elbow on *right*. A piece of cotton was put under the elbow to allow full extension of the elbow. (**b**) Data from one participant for full flexion at the wrist and full extension at the elbow on *left* and for full extension at the wrist and full flexion at the elbow on *right*. *Shading* shows the actual maintained posture of the arm and hand. *Solid lines* show the perceived starting position of the arm prior to cuff inflation and the perceived final position of the phantom arm at the end of the block [at the elbow (184° to 77°) and wrist (133° to 239°), at the elbow (53° to 123°) and wrist (208° to 135°)], respectively. The *dashed lines* are the perceived positions of the developing phantom at an intermediate time. Times are given relative to the time of cuff inflation. The direction of the perceived change is given by the *curved arrow*. (**c**) Data from ten participants for full flexion at the wrist and full extension at the elbow on *left* and for full extension at the wrist and full flexion at the elbow on *right* [at the elbow (189° to 86°) and wrist (136° to 192°), at the elbow (77° to 98°) and wrist (223° to 160°)] (Inui et al. 2012a)

wrist were measured using the left arm and hand attached to a goniometer system (DKH, Tokyo, Japan). The sensory test was also conducted by the same procedure and measurement as Sect. 2.2 of this chapter.

The major finding is that, despite the position of the arm and hand being fixed (in either flexion or extension), the perceived position of the elbow and wrist changed systematically in all participants after cuff inflation. As a new finding of the present study, if the actual wrist was fully extended while the actual elbow was fully flexed, then the perceived position of the wrist moved toward flexion, whereas that of the elbow moved toward extension. Conversely, if the actual wrist was fully flexed while the actual elbow was fully extended, then the wrist was perceived to extend, whereas the elbow was perceived to flex. Figure 2.8b, c shows schematics of data (on the left) of the perceived change in joint angle when the actual wrist was fully flexed while the actual elbow was fully extended and data (on the right) of the perceived change in joint angle when the actual wrist was fully extended while the actual elbow was fully flexed. These systematic changes in position did not involve arm and hand postures that were anatomically impossible, and the changes were in the same direction. There was a seamless perceived position for the arm and hand. The final position of the phantom limb further depended on its initial position.

Before this study was conducted, we anticipated that because changes in the perceived posture in two different directions were compensated for each other by interactions between biarticular muscles across the wrist and elbow, no change in the perceived posture at each joint would be experienced. For example, when the wrist was fully flexed while the elbow was fully extended, the flexors of the upper arm were stretched while the flexors of the forearm were contracted. It was anticipated that stretching of the flexors of the upper arm and contraction of the flexor of the forearm compensated for each other. Contrary to this expectation, however, the two perceived postures at the two joints obviously moved toward the opposite direction. Similar to the perceptions when the wrist and elbow were fixed in the same direction in our previous studies (Inui et al. 2011, 2012b), changes in the perceived posture at each joint depended on its initial position even when the two joints were fixed in the opposite direction.

The present study indicated that the flexor was stretched and the extensor was contracted at a joint so that the signal from the flexor was altered as afferent firing disappeared. Conversely, the flexor was contracted and the extensor was stretched at another joint so that disappearance of afferents altered the signal from the extensor. Although it was anticipated that there was no change in the perceived posture by interactions between biarticular muscles across the wrist and elbow, the disappearance of afferents in an extreme posture had a stronger effect on the perceived change than the interactions between biarticular muscles. Hence, the two perceived postures at the two joints moved toward opposite directions independently.

Inui et al. (2011) reported that, despite the position of the hand being fixed throughout the period of cuff inflation, the final perceived position of the hand was more flexed when the hand was held in extension than when it was held in flexion. In the present study as well as the previous study (Inui et al. 2012b), however, although there was no difference between extension and flexion at the wrist in the magnitude of change in the perceived posture, the extension of the elbow exhibited a markedly larger magnitude than the flexion of the joint. One plausible contributory mechanism is that elbow flexors and extensors are supplied by different nerves. Because the elbow flexor is innervated by the musculocutaneous nerve while the elbow extensor by the radial nerve, it is possible that the block of the radial nerve was less severe than that of the musculocutaneous nerve.

2.5 Effect of Visual Information on Perceived Posture of a Phantom Foot

Visual information plays an important role in the formation of body image. In particular, the role of the visual image of the body is well known by the observation that mirrors can induce synesthesia in a phantom limb (Hunter et al. 2003; Ramachandran and Rogers-Ramachandran 1996) and by the finding that illusions of ownership of a rubber hand can be elicited by synchronized visual and tactile stimuli in normal participants (Botvinick and Cohen 1998). In addition, Paqueron et al. (2003) reported the effect of sight following the description of the changes in body image induced by an anesthesia block. When most participants saw their anesthetized limb, the illusion of the position and posture of the anesthetized limbs was influenced by visual information, whereas the sensation of body shape alteration itself was not modified by the information. Silva et al. (2010) also reported that visual information caused a rapid superposition of the position of the phantom limb on the real posture of the anesthetized limb in all participants. Most participants felt the position illusion following the reintroduction of the visual mask again. However, Paqueron et al. (2003) and Silva et al. (2010) did not describe the effects of visual information on the body image quantitatively. Therefore, in this section the first aim of the study was to examine quantitative effects of visual information on the perceived posture following the investigation of the changes in body image induced by an ischemic block.

Our recent studies (Inui et al. 2011, 2012b) showed that if the fingers, wrist, and elbow were held straight before and during ischemic anesthesia, the final perceived hand and arm were bent at the fingers, wrist, and elbow and vice versa. The perceived posture of the body part may be determined by the muscle group stretched by that posture. The muscle group generates the higher level of proprioceptive discharge. Nerve block lowers the level of discharge and leads to a perceived change in posture. In contrast, the low discharge from the antagonist shortened by that posture plays no role in the observed changes. We hypothesized that if a joint is fixed in full extension or flexion before and during anesthesia, an experimental phantom joint moved toward the opposite position. Thus, to extend the systematic change of body image found by our previous studies, the second aim of this study was to track the change in perceived posture of the ankle and knee during ischemic anesthesia

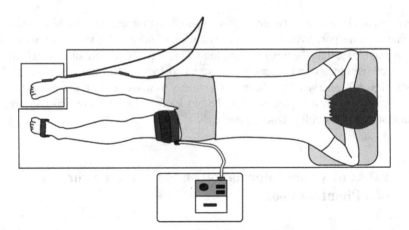

Fig. 2.9 Schematic of experimental setup. Participants lay comfortably prone on a table, and their right feet and legs were fixed by a tape with Velcro. Their heads and upper extremities were put on a pillow. A wide cuff was positioned on the right thigh and connected to the automatic tourniquet system. In the experiment with the right foot taped in full extension (see the *left panel* of Fig. 2.2a), the left foot was put on a separate table. Because the perceived posture of the right ankle was anticipated to become flexed in the experiment in full extension, the separate table was withdrawn when participants demonstrated the perceived posture of the right ankle using the left foot (Inui and Masumoto 2013)

of the thigh. Presumably, the flexed posture stretches knee extensors and ankle extensors, while the extended posture stretches ankle flexors. It is anticipated that the perceived posture of the flexed or extended knee and ankle joints is determined by the knee and ankle extensors or flexors stretched by that posture.

This study was performed on ten healthy male participants without any apparent neurological disorders (age range, 20–25 years). Participants lay comfortably on their stomach on a table with their closed eyes, and their right foot or leg was fixed by Velcro (see Fig. 2.9). This main study was conducted twice with all ten participants, once with the foot and leg fully extended (the left panel in Fig. 2.10a) and once with the foot and leg fully flexed (the right panel in Fig. 2.10a). Five participants were studied first in full extension, whereas the other five were studied first in full flexion.

A wide cuff (Zimmer, Dover, OH, USA) was positioned on the right thigh and connected to the automatic tourniquet system (Zimmer ATS 750) so that the cuff could be inflated to 250 mmHg in less than 1 s. Sensory tests included assessments of tactile sensation with von Frey hairs and the ability to detect a light touch with a cotton swab. Three sites for sensory testing were marked: the dorsum of the big toe, just proximal to the dorsum of the ankle, and the volar surface of the leg just distal to the knee. To measure perceived foot and leg postures during the development of the ischemic block, the participants demonstrated the perceived positions of the right ankle and knee using the left foot and leg attached to a flexible two-axis goniometer system (DKH, Tokyo, Japan). The participants were instructed to demonstrate the

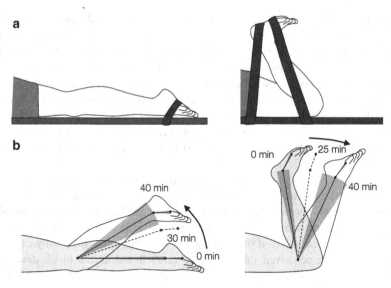

Fig. 2.10 Experimental setup (**a**) and perceived position of developing phantom during an ischemic block (**b**). (**a**) Full extension at the ankle and knee joints on *left* and full flexion at the ankle and knee joints on *right*. (**b**) Data from ten participants for the two versions of experiment one (ankle and knee extended on *left*, ankle and knee flexed on *right*). *Shaded legs* show the true maintained posture of the leg. *Solid lines* show the perceived starting position of the foot and leg prior to cuff inflation and the perceived final position of the phantom foot and leg at the end of the block. The *lines* represent mean values and they were given shadow outlines to indicate the standard errors. The *dashed lines* show the mean perceived position of the developing phantom at an intermediate time. Data at these times were determined from the regression fitted for each participant. Times are given relative to the time of cuff inflation. The direction of the perceived change is given by the *curved arrow*. If the ischemic block developed with the ankle and knee joints extended (*left*), then participants perceived that the ankle and knee joints became more flexed as the block develops [at the ankle (179° to 161°) and knee (180° to 150°)]. However, if the block was applied with the ankle and knee joints flexed (*right*), then the joints are perceived to become more extended as the block develops [at the ankle (142° to 167°) and knee (80° to 115°)] (Inui and Masumoto 2013)

perceived ankle and knee positions of their right foot and leg using their left foot and leg. The angle between the ankle joint and the little toe was measured as the perceived position of the ankle (see Fig. 2.10b).

The control measurements of touch and posture were taken by the same procedure as Sects. 2.1 and 2.2 of this chapter. In the main study the measurements of touch and posture were conducted at intervals of 5 min after inflation. To examine the effects of visual information on the perceived posture of both joints, the participants demonstrated the perceived posture of both joints using their left foot and leg after they looked at their right foot and leg at the end of the block.

Figure 2.10b shows schematics of the mean data from all ten participants who exhibited the magnitude of perceived change in joint angle for the two versions of the main experiment. The major finding is that, despite the position of the foot and leg being fixed (in either flexion or extension), after cuff inflation the perceived position of the ankle and knee changed systematically in all participants. If the actual foot and leg were fully extended, then the perceived position of the ankle and knee moved toward flexion. Conversely, if they were fully flexed, then their perceived position moved toward extension. This systematic change in position did not involve foot and leg postures that were anatomically impossible, and the changes were in the same direction. The final position of the phantom limb depended on its initial position. In addition, the foot and leg are perceived to move continuously with a seamless position. These results are consistent with the results of our previous studies (Inui et al. 2012a, b).

Figure 2.11a, b shows means and standard deviations of the absolute changes in perceived joint angles during control tests and ischemia. The control data showed that the participants correctly signaled that their foot and leg were flexed or extended with minimal adaptation for 10 min prior to cuff inflation. The data fitted to a sigmoid curve indicated that if the actual ankle and knee joints were fully extended (open circles), the perceived joints became more flexed as the block developed (ankle, $R^2 = 0.991$, $P < 0.0001$, $slope = 17.320$; knee, $R^2 = 0.996$, $P < 0.0001$, $slope = 31.349$). In contrast, both joints were perceived to become more extended (solid circles) if the block was applied with them in a fully flexed position (ankle, $R^2 = 0.985$, $P < 0.0001$, $slope = 23.310$; knee, $R^2 = 0.997$, $P < 0.0001$, $slope = 35.778$). The flexion of both joints exhibited a steeper slope than the extension of the joints.

Figure 2.11c, d showed means and standard deviations of differences between perceived joint angles during control tests and ischemia. When the ischemic block was applied with the ankle and knee joints fully extended (open circles), the difference between perceived joint angles during control tests and ischemia increased as the block developed. Similarly, the difference increased when the block was applied with them in a fully flexed position (solid circles). The magnitude of the perceived changes in both joints was larger when they were flexed than when they were extended. Although there was no significant difference between joints for the magnitude in flexion, the magnitude in the knee was larger than that in the ankle in extension. The interaction of joint and posture was thus significant.

As a new finding, Fig. 2.11c, d also showed means and standard deviations of differences between perceived joint angles during control tests and after the end of the experiment (vision). Although there was no significant difference between at 0 and 25 min after the block and at the moment of visualization for the magnitude of the perceived changes in both joints, the magnitude was larger at 30–40 min after the block than at the moment of visualization. The perceived posture of both joints thus returned nearly to the position at the earlier stage of the experiment after they looked at their right feet and legs at the end of the experiment.

Melzack and Bromage (1973) early found changes in position of a phantom arm after an anesthetic nerve block. When participants looked at the real limb, the phantom arm suddenly coincided or fused with the real arm in 65 % of the participants. The dissociation between phantom and real limbs occurred again when the eyes were closed. Paqueron et al. (2003) also showed that when most participants

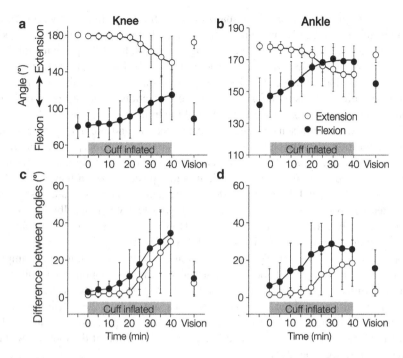

Fig. 2.11 Means and standard deviations of the absolute changes in perceived joint angles (**a, b**) and differences between perceived joint angles during control tests and ischemia (**c, d**). (**a**) Changes in perceived angle of the knee joint during ischemic block. Mean and standard error are obtained from ten participants for the knee. 180° represents a straight joint. Note data on visual information were calculated for seven participants in the extension version and six in the flexion version. When the ischemic block was applied with the joints extended (data in *open circles*), the knee appeared to become gradually more flexed as the block developed. In contrast, the knee was perceived to become steeply more extended when the block was applied with it in a flexed position (data in *solid circles*). The magnitude of the perceived change in the knee was less when the foot was extended than when it was flexed. The perceived posture of the knee returned nearly to the control position after the participants looked at their right feet and legs at the end of the experiment. (**b**) Changes in perceived angle of the ankle joint during development of an ischemic block. Perceived change of the ankle was less than that of the knee for both the extension and flexion versions. Similar to the knee, perceived posture of the ankle returned nearly to the control position after the participants looked at their right feet and legs at the end of the experiment. (**c**) Data for full extension (*open circles*) and flexion (*solid circles*) at the knee joint. Mean and standard deviation are obtained from ten participants for the knee. Note data on visual information were calculated for seven participants in the extension version and six in the flexion version. (**d**) Data for full extension and flexion at the ankle joint (Inui and Masumoto 2013)

saw or grasped their anesthetized limb, the illusion of its position and posture was influenced by visual information. Silva et al. (2010) similarly reported that visual information caused a rapid superposition of the position of the phantom limb on the real posture of the anesthetized limb in all participants. In the study of this chapter, although there was no significant difference between at 0 and 25 min after the block and at the moment of visualization for the magnitude of the perceived changes in both joints, the magnitude was larger at 30 min to 40 min after the block than at the moment of visualization. Although these results of the present study are consistent with the observations of previous studies (Melzack and Bromage 1973; Paqueron et al. 2003; Silva et al. 2010), the present study found for the first time the quantitative effect of visual information on the correction of illusory perception of a limb's position using ischemic anesthesia of the limb.

Essentially, the multisensory neurons in the parietal and premotor cortices that code the position of body parts favor visual input over proprioceptive input (for a review, Giummarra et al. 2007). The perceived positions of body parts regarding online proprioceptive input drift from their actual positions when they are hidden from view (Gross and Melzack 1978), in particular when they are resting or supported passively in a constant position (Lloyd 2007; Wann and Ibrahim 1992) or are passively moved (Botvinick 2004). However, visual input is not always more accurate than proprioceptive input. The classic finding that vision overrides proprioception only holds for specific conditions and spatial directions, for example, when the visual location of the body part falls within the region where bimodal visuo-proprioceptive neurons in the premotor and posterior parietal cortices encode spatial proximity around the body (peripersonal space) (Lloyd 2007). Similarly, the present study found that visual information corrected the perceived posture of the ankle and knee to the real posture because the visual location of the foot fell within peripersonal space around the foot.

Electrophysiological studies at a single neuron level indicate that an interaction between vision and proprioception may change cellular discharge (Duhamel et al. 1998; Fogassi et al. 1996; for review, Graziano and Botvinick 2002). For example, neurons in the premotor cortex may discharge not only in response to visual stimuli near a monkey's hand, but also to stimuli approaching a dummy monkey arm while the monkey's real arm is hidden from view (Graziano 1999). Neurons in parietal area 5 can also increase their tonic discharge when a dummy arm is presented that holds a posture similar to that of the real monkey's arm, which is hidden from view (Graziano et al. 2000). Further electrophysiological studies have reported that neurons in the ventral premotor cortex, ventral intraparietal area, posterior parietal cortex, and putamen show both visual and tactile receptive fields (for review, Maravita 2006). Each of the bimodal neurons codes a region of peripersonal visual space which is spatially aligned to the preferred somatosensory receptive field of that neuron. For example, the ventral intraparietal neurons with somatosensory receptive fields on the right upper face respond to stimuli presented to the right upper quadrant of the visual field (Duhamel et al. 1998), or a premotor neuron with a tactile receptive field on the arm or hand discharges in response to visual stimuli approaching that body (Graziano et al. 1994). These data suggest that somatosensory stimuli to the body are integrated with visual stimuli coming from extrapersonal space (Maravita 2006). The findings support the view that body image is constructed through the integration of inputs from different sensory modalities. Our results thus suggested that the effect of visual input on body image was associated with the interaction of sensory inputs found by electrophysiological studies at a single neuron level.

The magnitude of the perceived change at the ankle and knee in this study was less than that at the wrist and elbow in previous studies (Inui et al. 2012a, b). While the magnitude of the perceived change was 20°–30°at the ankle and 30°–40° at the knee, the magnitude reached 60° at the wrist and 80° at the elbow. Although the ankle has a smaller range of movement than the wrist, the range of the knee is similar to that of the elbow. Thus, the difference in the magnitude of the perceived changes between upper and lower limbs is not associated with the range of joint movement. When a cuff was positioned on the upper arm (Inui et al. 2012a, b), while large-diameter fiber function of the thumb of the hand and wrist, as assessed by cutaneous touch, was lost in all 20 cases of 10 participants at the end of an ischemia block, that of the elbow was not lost in 9 or 7 of 20 cases. In this study, the large-fiber function of the big toe of the foot and ankle was lost in all 20 cases at the end of the block, although that of the knee was not lost in 17 of 20 cases. In addition, when a cuff was positioned on the upper arm (Inui et al. 2012a), small-diameter fiber function of the thumb, as assessed by pain, was lost in all 20 cases of 10 participants at the end of the block, while that in the elbow was lost in only 5 of 20 cases. In this study, small-fiber function of the big toe was lost in 13 of 20 cases at the end of the block, although that of the knee was lost in only 1 of 20 cases. Although both studies used the same tourniquet method and interval, the upper limb had a deeper level of anesthesia than the lower limb because the lower limb has a larger volume of muscle and a thicker blood vessel than the upper limb. Thus, the difference in the magnitude of the perceived changes between upper and lower limbs may be associated with the depth of anesthesia.

In this study, the magnitude of the perceived changes in both joints was larger when they were flexed than when they were extended. These results exhibited the reverse observations of the upper limb in our previous studies (Inui et al. 2011, 2012a). Because the ankle extensor is innervated by the deep peroneal nerve, while the ankle flexor is innervated by the tibial nerve, one plausible contributory mechanism is that the block of the tibial nerve was less severe than that of the deep peroneal nerve. Similarly, because the knee extensor is innervated by the femoral nerve, while the knee flexor is innervated by the tibial and common peroneal nerves, it is possible that the block of the tibial and common peroneal nerves was less severe than that of the femoral nerve.

2.6 Change in Perceived Size of a Phantom Hand

Body posture is signaled by proprioceptive information provided by receptors in the muscle, skin, and joint (Proske and Gandevia 2012). However, there are no receptors to signal body size. In addition, although planning and executing movement require the brain to have some information about body size, the information does not need to be accessible to conscious perception. Longo and Haggard (2010) show that participants consciously perceive only the undistorted map about the body's metric properties. This suggests the dissociation between information about body size for motor control and that used for perception.

When the sensory input from a body part is removed by peripheral anesthesia, the part is not excised from perception. The part usually remains and is commonly described as swollen. As examples, local anesthesia of the thumb with a nerve block markedly increases the apparent size of the thumb (Gandevia and Phegan 1999), while a brachial plexus block produces illusory swelling of the arm (Paqueron et al. 2004a). Such distortions of body segment size have been linked to loss of input from small-diameter afferents (Calford and Tweedale 1991; Paqueron et al. 2003, 2004a). However, it is unclear whether the background firing of large-diameter afferents is involved. In this chapter we determine whether the perceived size of the hand increases following acute ischemic anesthesia of the upper arm. The form of anesthesia is chosen because ischemic anesthesia develops slowly with loss of conduction in large-diameter axons occurring before that in small-diameter axons.

The experiment was conducted by the same ten participants as Sect. 2.1 of this chapter. Perceived size of the hand was estimated by selection of a simple two-dimensional outline or template of the hand, in a neutral position, which best matched its "size." Templates of different sizes were randomly arranged on sheets (Gandevia and Phegan 1999; Longo and Haggard 2010). The range of magnifications was 50–190 % in 10 % increments, with 100 % being the size of a typical adult hand. Participants were asked to "select the template which best matches how big your hand feels or the perceived size of your hand." The participants chose one template from each of three sheets on each occasion. The participants gave their responses within 10–15 s. Template selections were made four times prior to cuff inflation and at intervals of 5 min after inflation.

Figure 2.12 shows changes in the perceived size of the hand. Across participants, there is a progressive increase in perceived size which begins shortly after cuff inflation. By the end of cuff inflation the mean increase is 34 ± 4 % (mean ± 95 % confidence interval).

Perceived swelling of a deafferented body part has been documented previously after complete anesthesia of the thumb produced by lignocaine (Gandevia and Phegan 1999). Paqueron et al. (2003) have also found illusions of swelling, elongation, or shortening of the deafferented limb produced by similar compounds. Previous studies (Calford and Tweedale 1991; Paqueron et al. 2003) have claimed that such perceptual distortion of shape and size of body parts differs clearly from posture illusion. On the contrary, the results of our study suggest that loss of large-fiber input from the hand is sufficient to initiate both the illusion of limb swelling and the alterations in perceived posture of the hand and wrist. However, whereas the mean increase for the perceived size of the hand is 34 % in our study, the increase for the perceived size of the finger is 60–70 % in the previous study (Gandevia and Phegan 1999). This increase ratio in the previous study is quite larger than the ratio in our study. It is unclear whether this difference is produced by different paralyzed peripheral nerves or by different sizes of the body part (hand and finger).

With progressive ischemic anesthesia, there is an accompanying illusion that the limb increases in size and becomes swollen. Perceived swelling of a deafferented body part has been documented previously after complete anesthesia of the thumb (Gandevia and Phegan 1999) and partial anesthesia of the face (Turker et al. 2005)

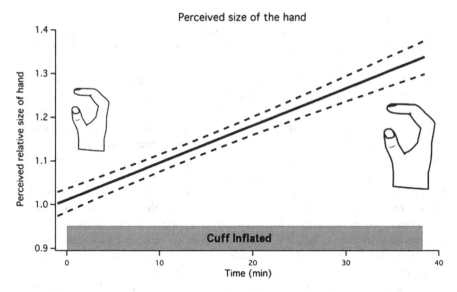

Fig. 2.12 The mean change in perceived size of the hand during an ischemic block. The *solid line* represents a regression line of the average of all ten participants with 95 % confidence interval (*dashed line*). The insets show two templates of the hand to illustrate the magnitude of the average increase in perceived size. Perceived size of the hand was estimated by selection of a simple two-dimensional outline or template of the hand, in a neutral position, which best matched its size. Templates of different sizes were randomly arranged on sheets. The range of magnifications was 50–190 % in 10 % increments, with 100 % being the size of a typical adult hand. Participants were asked to "select the template which best matches how big your hand feels or the perceived size of your hand." Participants chose one template from each of three sheets on each occasion. Participants gave their responses within 10–15 s. Template selections were made four times prior to cuff inflation and at intervals of 5 min after inflation (Modified from Inui et al. 2011)

produced by lignocaine, as well as during clinical nerve blocks affecting large regions of the limb produced by similar compounds (Paqueron et al. 2003, 2004a, b). Because the nerve blocks induced by lignocaine and by ischemia do not affect axonal function in the same way (see below and Grosskreutz et al. 1999), the present results suggest that loss of large-fiber input from the hand is sufficient to initiate both the illusion of limb swelling and the alterations in perceived posture of the hand and wrist. We note that some formal assessments of sensation show early changes prior to changes in the perceived posture of the limb (Silva et al. 2010) and that this is corroborated by subjective changes in limb size (Paqueron et al. 2004a).

The conventional view is that ischemia and focal pressure over nerves innervating the arm produce their effect differentially on fibers of different diameters (Gasser and Erlanger 1929; Mackenzie et al. 1975), with those of larger diameter being affected earlier than those of smaller diameter. Axonal diameter is not the only factor determining the effect of ischemia because populations of large-diameter sensory and motor axons have slightly different responses to ischemia (e.g., Mogyoros et al. 1997). Furthermore, the behavior of different classes of C fibers is not uniform (Serra et al. 1999). Our result showing resistance to prolonged ischemia of some C fibers, which signal heat pain, supports this view. By contrast, local anesthesia produced by lignocaine and related compounds preferentially disrupts the function of small-diameter axons, while large-diameter axons take longer to block (e.g., Paqueron et al. 2004a). The results from our method of ischemic block by cuff inflation on the upper arm show that it produced a progressive disruption of the afferent input from the forearm and hand. In all participants changes in sensory testing began first in the large-diameter cutaneous fibers, particularly those innervating the hand, and after 30–40 min, large-fiber function, assessed by cutaneous touch, was lost up to just below the cuff at elbow level. Smaller fibers in the Aδ range include the main population of cold fibers (Johnson et al. 1973; Mackenzie et al. 1975), and by the end of the ischemic period, we found a marked impairment of cold thresholds (by 5°–10°). In contrast, there was little impairment of heat pain, which indicates residual input from a population of C fibers subserving the sensation of pain to heating. These observations make it likely that both the distortions in perceived size of the limb and in its posture can be initiated and driven by loss of background input from large-diameter sensory axons. They do not reveal the additional role of classes of Aδ and C fibers in these illusions because it is difficult to abolish their background discharge selectively. Studies using local anesthetic block and subjective reports suggest that illusions of perceived size of the limb can also be initiated by changes occurring in small-diameter afferents (e.g., Paqueron et al. 2004b).

Such data indicate the brain uses afferent information to generate our sense of body size. Others have suggested that rapid cortical reorganization may be the mechanism involved (Calford and Tweedale 1991). In addition, perceived changes in body size may be related to removal of inhibitory background activity at several levels in the somatosensory pathway (Dykes and Lamour 1988; Dykes and Craig 1998). However, the size illusion is unable to illustrate exclusively from the cortical reorganization or the removal of the inhibitory activity because anesthesia of large body areas induces a change in perceived size in our study as well as the work of Paqueron et al. (2003). Moreover, if perceived body size were driven by receptive field enlargement, then perceived length should increase as much as perceived width. This is not the recent observation following local anesthesia by Walsh et al. (2015). They found that during anesthesia the finger is perceived to get up to 30 % fatter, but less than 3 % longer, suggesting a different organization not based on receptive field.

The width and length of limb segments are critical for skilled motor actions. While changes in limb length during adolescent are slow (Tanner 1962), those in the width of body segments are produced over short time frames. The difference in behavior of perceived width and length following anesthesia may be related to the likelihood of them changing rapidly. An increase in the perceived size of limb segments could be a mechanism to prevent injury, adding a safety margin to keep the body farther from possible source of injury. This suggests that the strategy negatively impacts motor performance because the representation of the body is inaccurate.

References

Aglioti S, DeSouza JF, Goodale MA (1995) Size-contrast illusions deceive the eye but not the hand. Curr Biol 5:679–685

Bergenheim M, Ribot–Ciscar E, Roll J–P (2000) Proprioceptive coding of two-dimensional limb movements in humans: I. Muscle spindle feedback during spatially oriented movements. Exp Brain Res 134:301–310

Botvinick M (2004) Probing the neural basis of body ownership. Science 305:782–783

Botvinick M, Cohen J (1998) Rubber hands 'feel' touch that eyes see. Nature 391:756

Brasil–Neto JP, Valls–Sole J, Pascual–Leone A, Cammarota A, Amassian VE, Cracco R, Maccabee P, Cracco J, Hallett M, Cohen LG (1993) Rapid modulation of human cortical motor outputs following ischaemic nerve block. Brain 116:511–525

Calford MB, Tweedale R (1991) C-fibres provide a source of masking inhibition to primary somatosensory cortex. Proc R Soc Lond B Biol Sci 243:269–275

Collins DF, Refshauge KM, Todd G, Gandevia SC (2005) Cutaneous receptors contribute to kinesthesia at the index finger, elbow, and knee. J Neurophysiol 94:1699–1706

Duhamel JR, Colby CL, Goldberg ME (1998) Ventral intraparietal area of the macaque: congruent visual and somatic response properties. J Neurophysiol 79:126–136

Dykes RW, Craig AD (1998) Control of size and excitability of mechanosensory receptive fields in dorsal column nuclei by homolateral dorsal horn neurons. J Neurophysiol 80:120–129

Dykes RW, Lamour Y (1988) An electrophysiological laminar analysis of single somatosensory neurons in partially deafferented rat hindlimb granular cortex subsequent to transection of the sciatic nerve. Brain Res 449:1–17

Edin BB (1992) Quantitative analysis of static strain sensitivity in human mechanoreceptors from hairy skin. J Neurophysiol 67:1105–1113

Edin BB, Johansson N (1995) Skin strain patterns provide kinaesthetic information to the human central nervous system. J Physiol 487:243–251

Flor H, Elbert T, Knetch S, Wienbruch C, Pantev C, Birbaumer N, Larbig W, Taub E (1995) Phantom limb as a perceptual correlate of cortical reorganization following arm amputation. Nature 375:482–484

Flor H, Nikolajsen L, Jensen TS (2006) Phantom limb pain: a case of maladaptive CNS plasticity? Nat Rev Neurosci 7:873–881

Fogassi L, Gallese V, Fadiga L, Luppino, Matelli M, Rizzolatti G (1996) Coding of peripersonal space in inferior premotor cortex (area F4). J Neurophysiol 76:141–157

Gandevia SC (1985) Illusory movements produced by electrical stimulation of low-threshold muscle afferents from the hand. Brain 108:965–981

Gandevia SC, Phegan CM (1999) Perceptual distortions of the human body image produced by local anaesthesia, pain and cutaneous stimulation. J Physiol 514:609–616

Gandevia SC, Smith JL, Crawford M, Proske U, Taylor JL (2006) Motor commands contribute to human position sense. J Physiol 571:703–710

Gasser HS, Erlanger J (1929) Role of fiber size in establishment of nerve block by pressure or cocaine. Am J Physiol 88:581–591

Gentili ME, Verton C, Kinirons B, Bonnet F (2002) Clinical perception of phantom limb sensation in patients with brachial plexus block. Eur J Anaesth 19:105–108

Giummarra MJ, Gibson SJ, Georgiou–Karistianis N, Bradshaw JL (2007) Central mechanisms in phantom limb perception: the past, present and future. Brain Res Rev 54:219–232

Goodwin GM, McCloskey DI, Matthews PB (1972) The contribution of muscle afferents to kinaesthesia shown by vibration induced illusions of movement and by the effects of paralysing joint afferents. Brain 95:705–748

Graziano MSA (1999) Where is my arm? The relative role of vision and proprioception in the neuronal representation of limb position. Proc Natl Acad Sci U S A 96:10418–10421

Graziano MSA, Botvinick MM (2002) How the brain represents the body: insights from neuro-physiology and psychology. In: Prinz W, Hommel B (eds) Common mechanisms in perception and action, vol XIX, Attention and Performance. Oxford University Press, New York, pp 136–157

Graziano MSA, Yap GS, Gross CG (1994) Coding of visual space by premotor neurons. Science 266:1054–1057

Graziano MSA, Cooke DF, Taylor CS (2000) Coding the location of the arm by sight. Science 290:1782–1786

Gross Y, Melzack R (1978) Body image: dissociation of real and perceived limbs by pressure-cuff ischemia. Exp Neurol 61:680–688

Grosskreutz J, Lin C, Mogyoros I, Burke D (1999) Changes in excitability indices of cutaneous afferents produced by ischaemia in human subjects. J Physiol 518:301–314

Henderson WR, Smyth GE (1948) Phantom limbs. J Neurol Neurosurg Psychiatry 11:88–112

Hunter JP, Katz J, Davis KD (2003) The effect of tactile and visual sensory inputs on phantom limb awareness. Brain 126:579–589

Inui N, Masumoto J (2013) Effects of visual information on perceived posture of an experimental phantom foot. Exp Brain Res 226:487–494

Inui N, Walsh LD, Taylor JL, Gandevia SC (2011) Dynamic changes in the perceived posture of the hand during ischaemic anaesthesia of the arm. J Physiol 589:5775–5784

Inui N, Masumoto J, Beppu T, Shiokawa Y, Akitsu H (2012a) Loss of large-diameter nerve sensory input changes the perceived posture. Exp Brain Res 221:369–375

Inui N, Masumoto J, Ueda Y, Ide K (2012b) Systematic changes in the perceived posture of the wrist and elbow during formation of a phantom hand and arm. Exp Brain Res 218:487–494

Johnson KO, Darian-Smith I, LaMotte C (1973) Peripheral neural determinants of temperature discrimination in man: a correlative study of responses to cooling skin. J Neurophysiol 36:347–370

Kammers MP, van der Ham IJ, Dijkerman HC (2006) Dissociating body representations in healthy individuals: differential effects of a kinaesthetic illusion on perception and action. Neuropsychologia 44:2430–2436

Levy LM, Ziemann U, Chen R, Cohen LG (2002) Rapid modulation of GABA in sensorimotor cortex induced by acute deafferentation. Ann Neurol 52:755–761

Lloyd DM (2007) Spatial limits on referred touch to an alien limb may reflect boundaries of visuo-tactile peripersonal space surrounding the hand. Brain Cogn 64:104–109

Longo MR, Haggard P (2010) An implicit body representation underlying human position sense. Proc Natl Acad Sci U S A 107:11727–11732

Longo MR, Azanon E, Haggard P (2010) More than skin deep: body representation beyond primary somatosensory cortex. Neuropsychologia 48:655–668

Mackenzie RA, Burke D, Skuse NF, Lethlean AK (1975) Fibre function and perception during cutaneous nerve block. J Neurol Neurosurg Psychiatry 38:865–873

Maravita A (2006) From "body in the brain" to "body in space": sensory and intentional components of body representation. In: Knoblich G, Thornton IM, Grosjean M, Shiffrar M (eds) Human body perception from the inside out. Oxford University Press, New York, pp 65–88

Melzack R, Bromage PR (1973) Experimental phantom limbs. Exp Neurol 39:261–269

Mercier C, Reilly KT, Vargas CD, Aballea A, Sirgu A (2006) Mapping phantom movement representations in the motor cortex of amputees. Brain 129:2202–2210

Mogyoros I, Kiernan MC, Burke D, Bostock H (1997) Excitability changes in human sensory and motor axons during hyperventilation and ischaemia. Brain 120:317–325

Paqueron X, Leguen M, Rosenthal D, Coriat P, Willer JC, Danziger N (2003) The phenomenology of body image distortions induced by regional anaesthesia. Brain 126:702–712

Paqueron X, Gentili ME, Willer JC, Coriat P, Riou B (2004a) Time sequence of sensory changes after upper extremity block: swelling sensation is an early and accurate predictor of success. Anesthesiology 101:162–168

Paqueron X, Leguen M, Gentili ME, Riou B, Coriat P, Willer JC (2004b) Influence of sensory and proprioceptive impairment on the development of phantom limb syndrome during regional anesthesia. Anesthesiology 100:979–986

Pascual-Leone A, Valls-Sole J, Wassermann EM, Hallett M (1994) Responses to rapid-rate transcranial stimulation of the human motor cortex. Brain 117:847–858

Proske U, Gandevia SC (2012) The proprioceptive senses: their roles in signaling body shape, body position and movement, and muscle force. Physiol Rev 92:1651–1697

Ramachandran VS, Rogers-Ramachandran D (1996) Synaesthesia in phantom limbs induced with mirrors. Proc R Soc Lond B 263:377–386

Ribot-Ciscarb E, Roll JP (1998) Ago-antagonist muscle spindle inputs contribute together to joint movement coding in man. Brain Res 791:167–176

Ridding MC, Rothwell JC (1995) Reorganization in human motor cortex. Can J Physiol Pharmacol 73:218–222

Roll JP, Vedel JP (1982) Kinaesthetic role of muscle afferents in man, studied by tendon vibration and microneurography. Exp Brain Res 47:177–190

Seizova-Cajic T, Smith JL, Taylor JL, Gandevia SC (2007) Proprioceptive movement illusions due to prolonged stimulation: reversals and aftereffects. PLoS ONE 2, e1037

Serra J, Campero M, Ochoa J, Bostock H (1999) Activity-dependent slowing of conduction differentiates functional subtypes of C fibres innervating human skin. J Physiol 515:799–811

Silva S, Bataille B, Jucla M, Minville V, Samii K, Fourcade O, Démonet J-F, Loubinoux I (2010) Temporal analysis of regional anaesthesia-induced sensorimotor dysfunction: a model for understanding phantom limb. Br J Anaesth 105:208–213

Sirigu A, Grafman J, Bressler K, Sunderland T (1991) Multiple representations contribute to body knowledge processing. Evidence from a case of autotopagnosia. Brain 114:629–642

Tanner JM (1962) Growth at adolescence, 2nd edn. Blackwell Scientific Publications, Oxford

Turker KS, Yeo PL, Gandevia SC (2005) Perceptual distortion of face by local anaesthesia of the human lips and teeth. Exp Brain Res 165:37–43

Vallence AM, Reilly K, Hammond G (2012) Excitability of intracortical inhibitory and facilitatory circuits during ischemic nerve block. Restor Neurol Neurosci 30:345–354

Walsh LD, Gandevia SC, Taylor JL (2010) Illusory movements of a phantom hand grade with the duration and magnitude of motor commands. J Physiol 588:1269–1280

Walsh LD, Hoad D, Rothwell JC, Gandevia SC, Haggard P (2015) Anaesthesia changes perceived finger width but not finger length. Exp Brain Res 233:1761–1771

Wann JP, Ibrahim SF (1992) Does limb proprioception drift? Exp Brain Res 91:162–166

Ziemann U, Corwell B, Cohen LG (1998) Modulation of plasticity in human motor cortex after forearm ischemic nerve block. J Neurosci 18:1115–1123

Chapter 3
A New Type of Hand–Object Illusion

Abstract In Sect. (2.3) of Chap. 2, we previously showed that if the actual arm and hand were fully extended during ischemic anesthesia, then the perceived position of the elbow and wrist moved toward flexion. In this chapter, the study thus examined whether the wrist and elbow were perceived as flexed when a stick was fixed to the hand while the joints were extended during the ischemic block. Ten healthy participants lay on their back on a bed with their eyes closed, and a stick was fixed to their right hand. Surprisingly, while the perceived position of the wrist and elbow moved toward flexion from 10 to 40 min after the block, the stick fixed to the hand was also perceived as having moved toward flexion from 10 to 20 min after the block. Such coupling the change in the perceived stick position with the change in body image suggests a new type of hand–object illusion.

Keywords Body image • Hand–object illusion • Ischemic anesthesia

The ability to manipulate objects, balls, or sticks is one of the most remarkable perceptual and motor skills of the human species. This issue has been previously addressed from a motor control perspective (Jeannerod et al. 1995; Ehrsson et al. 2000), but we have little information about tactile and kinesthetic signals from the hand for manual performance (Rothwell et al. 1982; Johansson and Westling 1984; Jones and Lederman 2006). In particular, information about the spatial relationship between the hand and a handheld object when we manipulate the object is important.

Naito and Ehrsson (2006) examined the perceptual aspect of interaction between hand and object using a kinesthetic illusion (Goodwin et al. 1972). Right-handed participants touched a ball with the palm of their right or left hand while the tendon of the same wrist extensor muscle was vibrated. The wrist was perceived as flexing, and the ball was also perceived as moving along with the hand (hand–object illusion), while the left inferior parietal lobule was activated during hand–object illusions with both the right and left hands. Naito and Ehrsson (2006) reported that an external object is linked with our own hand in the left parietal lobule when the object becomes incorporated into our body image (also see Holmes et al. 2008).

This chapter examined the perceptual changes of interaction between hand and object using an acute ischemic block of peripheral nerves to the upper arm to create an experimental phantom hand (Gross and Melzack 1978). Our recent studies (Inui et al. 2011, 2012a, b; Inui and Masumoto 2013) took an experimental

© The Author(s) 2016   37
N. Inui, *Systematic Changes in Body Image Following Formation of Phantom Limbs*,
SpringerBriefs in Biology, DOI 10.1007/978-981-10-1460-4_3

phantom to mean a body part that was perceived although the large-diameter sensory input from that body part was completely gone. The studies showed that if the fingers, wrist, elbow, ankle, and knee were held straight before and during ischemic anesthesia, the perceived limb was bent at the joint and vice versa at the end of anesthesia. It is anticipated that if a joint is fixed in extension during the anesthesia, the final joint is perceived as flexed. Thus, in this chapter, the study demonstrates that if a stick is fixed to the hand as an extension of the wrist and elbow during anesthesia, the joints are perceived as flexing and the stick is also perceived as moving along with the hand (hand–object illusion). The study also supports the idea that this hand–object illusion depends on the central integration of kinesthetic and haptic information.

This study was performed on ten healthy male participants without any apparent neurological disorders (age range, 20–25 years). Participants lay comfortably on their backs on a bed with their eyes closed, and a stick (length, 50 cm; weight, 150 g) made by polyvinyl chloride was fixed to their right hand by a triangle bandage. Their right hand was also fixed by a Velcro (see Fig. 3.1a). This main study was conducted in all ten participants with the wrist and elbow extended. A piece of cotton was put under the elbow to fully extend the elbow. However, the wrist was unable to extend fully because the stick was fixed to the right hand by a triangle bandage.

A wide cuff (Zimmer, Dover, OH, USA) was positioned on the right upper arm and connected to the automatic tourniquet system (Zimmer ATS 750) so that the cuff could be inflated to 250 mmHg in less than 1 s. In the same procedure as Sect. (2.3) of Chap. 2, sensory tests included assessments of tactile sensation with von Frey hairs and the ability to detect a light touch with a cotton swab. To measure perceived wrist and elbow postures and stick positions during the development of the ischemic block, the participants were instructed to demonstrate the perceived wrist, elbow, and stick positions using the left wrist and elbow with the stick. To determine the time at which perception of the stick in the phantom hand was lost, the participants verbally reported whether they felt that they had a stick in their hand during the block or not. In addition, the participants were asked to report the perceived length of the stick. They verbally reported whether the length of the stick during the ischemic block was perceived shorter or longer than that before the block. Because main data on this study were based on self-reports, expectations of the participants could greatly contribute to reported results. Thus, the participants were not told the aim of the study by experimenters.

In pretest trials before inflation of the cuff, no change in the perceived position was confirmed when the hand and arm were maintained in the test posture. Pretest trials were carried out four times by the participants with their eyes closed for 10 min. Pretest measurements of touch and posture were taken at the start of the control trial and intervals of 2 min and 30 s. In test trials over a period of 40 min, the measurements of touch and posture were conducted at intervals of 5 min after inflation. Control trials were carried out in ten participants over a period of 50 min. The participant's perceived position of their intact wrist and elbow was measured by the same procedure as used during pretest and test trials when the hand with the stick was maintained in the test posture with the wrist and elbow held extended.

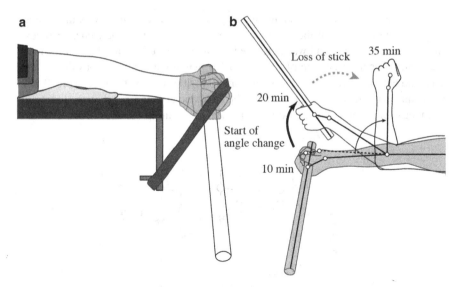

Fig. 3.1 Perceived positions of the wrist, elbow, and stick during control and test trials. (**a**) Means and standard deviations of the perceived angle during control and test trials for different time periods. During control trials, while open circles show data at the wrist joint, open triangles show those at the elbow joint. During test trials, while *solid circles* show data at the wrist joint, *solid triangles* show those at the elbow joint. Data on the left are means of the pretest measurement at each joint. Although data of mean and standard deviation during test trials were obtained from ten participants until 20 min, the data at 25, 30, 35, and 40 min are the mean for nine, eight, five, and two, respectively. (**b**) Means and standard deviations of two perceived stick positions during test trials for different time periods. While *open circles* show data at the stick position to the wrist, *solid circles* show those at the stick position to the *horizontal line*. Although data were obtained from ten participants until 15 min, the data at 20 and 25 min consisted of 4 and 2, respectively. Data on the left are means of the pretest measurement. (**c**) Median and interquartile range were plotted for start of changes in wrist and elbow joint angles and perceived loss of the stick (Inui and Masumoto 2015, Redrawn with permission from the Journal of Motor Behavior)

Tracking was performed by an infrared tracking system (OptiTrack V120, SenzTech, Beijing, China; pixel size, $6 \times 6 \, \mu m$; imager size, 4.5×2.88 mm; imager resolution, 640×480). While the positions of the wrist, elbow, and stick were tracked at 120 Hz by the infrared tracking system, the output of the system was recorded by a personal computer. Markers were attached on the dorsum of the proximal index, dorsum of the wrist, dorsum of the elbow, and opposite ends of the stick. The stick position was measured as "the stick angle to the wrist" from the tip of the stick, the dorsum of the wrist, and the dorsum of the elbow. The other stick position measured as "the stick angle to the horizontal line" was deemed to be at 180° with the wrist in line with the upper arm. The perceived stick angle to the horizontal line thus equals the perceived elbow angle plus the perceived wrist angle. We calculated the angle between two markers relative to the horizontal axis. This setup allowed us to compute the positions of the wrist and elbow with sub-0.5-mm precision.

This study found that the change in the perceived stick position was induced as the perceived postures of the wrist and elbow changed during the ischemic block. The stick fixed to the hand was perceived as having moved toward flexion from 10 to 20 min after the ischemic block (Fig. 3.1b), while the perceived position of the wrist and elbow moved toward flexion from 10 to 40 min after the block (Fig. 3.2a). The changes in the perceived posture of the hand and arm depended on the fading of cutaneous sensation from large-diameter nerve fiber during the anesthesia. The loss

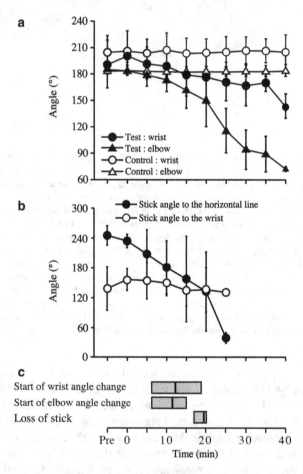

Fig. 3.2 Changes in sensation during the development of the ischemic block. (**a**) Changes in tactile threshold (von Frey) over three sites for sensory testing. Data at each time are means of ten participants. However, while data for the hand at 30 min are the means for three participants, the data at 35 min for the wrist and elbow are the means for four. Because data at 35 min for the hand and at 40 min for the wrist and elbow were for one participant, the data were not plotted. Data on left are means of the pretest measurement for each site. (**b**) Timing of changes in touch sensations. Data are shown for assessments made over the base of the thumb (*upper panel*), just proximal to the wrist (*middle panel*), and just proximal to the elbow (*lower panel*). Median and interquartile range are plotted for the start of changes in tactile threshold and the losses of light touch (Inui and Masumoto 2015, Redrawn with permission from the Journal of Motor Behavior)

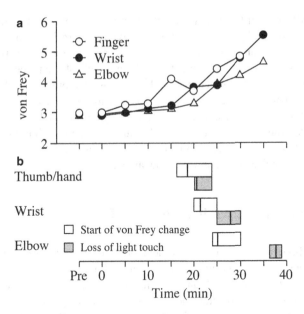

Fig. 3.3 Experimental setup and schematics of perception of change in joint angle. (**a**) Experimental setup. Participants lay comfortably on their *back* on a bed with their eyes closed, and a stick was fixed to their right hand by a triangle bandage. Their right hand was also fixed by Velcro tape. A piece of cotton was put under the elbow to fully extend the elbow. A wide cuff was positioned on the right upper arm and connected to the automatic tourniquet system. (**b**) Schematics of data of perceived change in joint angle and stick position when the actual arm and hand were fully extended. While data were obtained from ten participants until 15 min after inflation, the data consisted of four at 20 min after inflation. Shading shows the actual maintained posture of the arm and hand. *Solid lines* show the perceived starting position of the arm prior to cuff inflation, perceived final position of the phantom arm at the end of the block, and perceived final position of the stick. The *dashed lines* are perceived positions of the developing phantom at an intermediate time. Times are given relative to the time of cuff inflation. The direction of the perceived change is shown by the *curved arrow* (Inui and Masumoto 2015, Redrawn with permission from the Journal of Motor Behavior)

of perception of the stick in the hand occurred between the initial impairment and the complete loss of touch sensation (Figs. 3.2c and 3.3b). The hand and forearm with the stick were thus perceived as flexed between the start of the perceived change in the wrist and elbow and the complete loss of touch in the hand (Figs. 3.2c and 3.3b).

3.1 Development of Phantom Wrist and Elbow

The major finding of this study is that, although the hand and arm were fixed in extension, the perceived position of the wrist and elbow changed systematically in all participants after cuff inflation. Similar to the results of our previous studies (Inui et al. 2011, 2012a, b), if the actual hand and arm were extended, then the perceived position of the wrist and elbow moved toward flexion. Figure 3.1b shows schematics of data of perceived change in joint angle in the extension of the actual arm and hand.

Figure 3.2a shows the means and standard deviations of the perceived angle for different time intervals during control and test trials. Although both the wrist and elbow angles did not differ across time intervals during control trials, they decreased significantly during test trials. Then, the elbow angle decreased more steeply than the wrist angle. The results indicated that because the hand holding the stick could not be fully extended at the wrist while the elbow was fully extended, the wrist angle did not change more steeply than the elbow angle.

3.2 Perception of Stick in a Phantom Hand

Figure 3.2b shows means and standard deviations of two perceived stick positions during test trials for different time periods. Although the stick angle to the wrist did not differ across times, the stick angle to the horizontal line differed across times. The stick angle to the horizontal line decreased more steeply than the stick angle to the wrist because the elbow angle decreased more steeply than the wrist angle (Fig. 3.2a).

Figure 3.2c shows that while perception of a change in the wrist angle began at 12.0 [6.0–19.0] (median [IQR]) min, the change in the elbow angle began at 11.0 [6.0–15.0] min. The results also indicated that because the hand with the stick could not be fully extended at the wrist, the change in the wrist angle was slower than that in the elbow angle. The loss of perception of the stick in the hand began at 19.5 [16.0–20.0] min after the block. The stick fixed to the hand was thus perceived to move toward flexion from about 10 to 20 min after the block, while the perceived position of the wrist and elbow moved toward flexion from 10 to 40 min after the block.

Eight participants of ten reported that the stick in the hand was perceived as the shortened stick as ischemic anesthesia progressed although the perceived length of the stick cannot be measured quantitatively. They felt that there was something in the hand just before the loss of perception of the stick.

3.3 Changes in Sensation Following Cuff Inflation

There was a progressive degradation in sensation during development of the ischemic block. Figure 3.3a shows the sensory loss during ischemia began distally similar to our previous study in Sect. (2.2) of Chap. 2.

Figure 3.3b indicates that the loss of light touch followed the end of changes in the tactile threshold over the three testing sites. At the end of ischemia, while cutaneous touch sensation of the hand and wrist was lost by all of ten participants, that of the elbow was lost by only one of ten participants. The initial impairment (an increase in tactile threshold) began at 19.0 [16.0–24.0] min for the hand and at 25.0 [24.0–30.0] min for the elbow. Complete loss of detection of light touch followed the same pattern, again occurring earlier in the hand (21.0 [20.5–23.5] min) than in the elbow (37.5 [36.0–39.0] min).

3.4 A New Type of Hand–Object Illusion

The change in the perceived stick position was induced as the perceived postures of the wrist and elbow changed during the ischemic block (Figs. 3.1b and 3.2a). Because the elbow was fully extended while the hand holding the stick could not be fully extended at the wrist, the perceived elbow angle decreased more steeply than the perceived wrist angle. The perceived stick angle to the horizontal line equals the perceived elbow angle plus the perceived wrist angle. The perceived stick angle to the horizontal line thus decreased more steeply than the perceived stick angle to the wrist. Such coupling the change in the perceived stick position with the change in body image suggests a new type of hand–object illusion.

Related to the hand–object illusion, Naito and Ehrsson (2006) examined which areas of the brain are associated with the somatic sensation of hand–object interactive movement using functional magnetic resonance imaging. They indicated that an external object was linked with one's own hand in the left parietal lobule when the object became incorporated into one's body image. Similarly, the present study could speculate that the left parietal cortex linked the stick with the phantom hand, while the stick became incorporated into the body image. The skin contact between the palm and the object may stimulate slow-adapting receptors in the skin (Johansson and Vallbo 1983). Thus, the hand–object interaction illusion depends on the central integration of kinesthetic and haptic information (Gibson 1962), suggesting that the central integration underlies the stereognostic perception of the object (also see de Vignemont 2010).

3.5 Relationships Between Changes in Sensation and Illusion

In the study of this chapter, the elbow angle change was statistically different from pretest trials at 15–40 min after the block. Participants had lost perception of the stick at 20 min after the block. Therefore, for the first 20 min where the participants retained perception of the stick, the elbow moved only for 5 min.

In a median and IQR (Fig. 3.3b), on the other hand, the loss of the stick perceived in the hand began at 19.5 [16.0–20.0] min after the block. For the hand, while the initial impairment of touch sensation began at 19.0 [16.0–24.0] min, complete loss of detection of touch began at 21.0 [20.5–23.5] min. Although the loss of perception of the stick in the hand nearly coincided with the initial impairment of touch sensation, to be accurate, the loss of perception of the stick occurred between the initial impairment and the complete loss of touch sensation. The complete loss of touch perception resulted in the complete loss of perception of the stick in the hand. While perception of a change in the wrist angle began at 12.0 [6.0–19.0] min, the perception of a change in the elbow angle began at 11.0 [6.0–15.0] min. The right hand and forearm with the stick were perceived as flexed between the start of the perceived change in the wrist and elbow and the complete loss of touch in the hand. For this short period (about 10 min), the right hand and forearm constituted an incomplete phantom limb because of incomplete loss of detection of touch in the hand, wrist, and elbow. Brasil-Neo et al. (1993) have reported that corticospinal excitability increases in muscles proximal to an ischemic nerve block within 7–8 min following the block. The starting time of the increase in corticospinal excitability approximately corresponds to that of a change in the wrist and elbow angles in the present study. The existing cortical circuits might be disinhibited from this time after the block. This new finding that the incomplete phantom hand with the stick is perceived as having moved suggests a new type of hand–object illusion on the basis of the rapid changes of the cortical circuits.

However, because participants never actually use the stick, it was not perceived as a tool but a mere object held by the hand. In future studies, to demonstrate that the stick as a tool is incorporated in body representations, we plan the effect of ischemic block on the perception of a stick held by the hand after actually using the stick.

References

Brasil-Neto JP, Valls-Sole J, Pascual-Leone A, Cammarota A, Amassian VE, Cracco R, Maccabee P, Cracco J, Hallett M, Cohen LG (1993) Rapid modulation of human cortical motor outputs following ischaemic nerve block. Brain 116:511–525

de Vignemont F (2010) Body schema and body image – pros and cons. Neuropsychologia 48:669–680

Ehrsson HH, Fagergren E, Jonsson T, Westling G, Johansson RS, Forssberg H (2000) Cortical activity in precision versus power-grip tasks: an fMRI study. J Neurophysiol 83:528–536

Gibson JJ (1962) Observations on active touch. Psychol Rev 69:477–490

Goodwin GM, McCloskey DI, Matthews PB (1972) The contribution of muscle afferents to kinaesthesia shown by vibration induced illusions of movement and by the effects of paralysing joint afferents. Brain 95:705–748

Gross Y, Melzack R (1978) Body image: dissociation of real and perceived limbs by pressure-cuff ischemia. Exp Neurol 61:680–688

Holmes NP, Spence C, Hansen PC, Mackay CE, Calvert GA (2008) The multisensory attentional consequences of tool use: a functional magnetic resonance imaging study. PLoS One 3:e3502

Inui N, Masumoto J (2013) Effects of visual information on perceived posture of an experimental phantom foot. Exp Brain Res 226:487–494

Inui N, Masumoto J (2015) Perceptual changes of interaction between hand and object in an experimental phantom hand. J Mot Behav 47:81–88

Inui N, Walsh LD, Taylor JL, Gandevia SC (2011) Dynamic changes in the perceived posture of the hand during ischaemic anaesthesia of the arm. J Physiol 589:5775–5784

Inui N, Masumoto J, Beppu T, Shiokawa Y, Akitsu H (2012a) Loss of large-diameter nerve sensory input changes the perceived posture. Exp Brain Res 221:369–375

Inui N, Masumoto J, Ueda Y, Ide K (2012b) Systematic changes in the perceived posture of the wrist and elbow during formation of a phantom hand and arm. Exp Brain Res 218:487–494

Jeannerod M, Arbib MA, Rizzolatti G, Sakata H (1995) Grasping objects: the cortical mechanisms of visuomotor transformation. Trends Neurosci 18:314–320

Johansson RS, Vallbo AB (1983) Tactile sensory coding in the glabrous skin of the human hand. Trends Neurosci 6:27–32

Johansson RS, Westling G (1984) Roles of glabrous skin receptors and sensorimotor memory in automatic control of precision grip when lifting rougher or more slippery objects. Exp Brain Res 56:550–564

Jones LA, Lederman SJ (2006) Active haptic sensing. In: Jones LA, Lederman SJ (eds) Human hand function. Oxford University Press, New York, pp 75–99

Naito E, Ehrsson HH (2006) Somatic sensation of hand-object interactive movement is associated with activity in the left inferior parietal cortex. J Neurosci 267:3783–3790

Rothwell JC, Traub MM, Day BL, Obeso JA, Thomas PK, Marsden CD (1982) Manual motor performance in a deafferented man. Brain 105:515–542

Chapter 4
Visual and Proprioceptive Adaptation of Arm Position in a Virtual Environment

Abstract This chapter examined the resolution of a discrepancy between visual and proprioceptive estimates of arm position. Ten participants fixed their right shoulder at 0°, 30°, or 60° of transverse adduction, while they viewed a video on a head-mounted display that showed their right arm extended in front of the trunk for 30 min. The perceived arm position more closely approached the seen arm position on the display as the difference between the actual and visually displayed arm positions increased. This indicates that proprioceptive estimate adapted to match the visual estimate. By contrast, in the extreme case of a 90° discrepancy, the seen arm position on the display was very gradually perceived as approaching the actual arm position. This indicates that visual estimate of the arm position adapted to match the proprioceptive estimate. In addition, the magnitude of changes in sensory estimates was larger for proprioception (20 %) than for vision (<10 %).

Keywords Body image • Vision • Proprioception • Virtual environment

We both see and feel the position of our limbs. In daily life, although Smeets et al. (2006) suggest that visual and proprioceptive estimates are not aligned, one generally has the impression that one feels the limb is at same position at which it is seen. However, these sensory signals conflict when looking through a prism (Harris 1963; Martin et al. 1996; Redding and Wallace 1996) or using a virtual reality setup (van Beers et al. 2002).

This chapter examined the resolution of a discrepancy between visually observed and proprioceptively felt arm position using a two-arm position-matching task in which participants were presented with a visually displayed arm position on a 3D head-mounted display. This study thus examined perceptual rather than sensori-motor adaptation. Warren and Cleaves (1971) studied the relative dependence on vision and proprioception over a short period when these modalities gave discrepant information about the azimuth position of a target. Both bias effects decreased with increasing levels of discrepancy. In contrast to their study, the present study uses very large discrepancies between actual and visually displayed arm positions (up to 90°) and examines how the brain resolved the discrepancies between sensory modalities over a long period (half an hour).

© The Author(s) 2016

N. Inui, *Systematic Changes in Body Image Following Formation of Phantom Limbs*, SpringerBriefs in Biology, DOI 10.1007/978-981-10-1460-4_4

In addition, vision has been assumed to dominate any discrepancy between sensory modalities in most previous studies (for review, see Welch and Warren (1986)). However, the classic result should hold only for specific spatial directions (van Beers et al. 1999). Proprioception is more precise in depth than in azimuth, whereas vision is more precise in azimuth than in depth (van Beers et al. 1998, 2002). Because the modality weighted most heavily will adapt least (Ghahramani et al. 1997), proprioceptive adaptation is larger in azimuth than in depth (van Beers et al. 2002). In this chapter, we hypothesized that effects of vision on proprioception would be larger than those of proprioception on vision because the discrepancy between the actual and the visually displayed arm position was in azimuth. Therefore, the first aim of this chapter was to demonstrate the adaptation of proprioceptive estimates of actual arm position to match the visual estimates of the visually displayed arm position. The second aim was to demonstrate the adaptation of visual estimates to match the proprioceptive estimates.

This study was performed using ten healthy male participants without any apparent neurological disorders (mean age = 22.6 years, standard deviation (SD) = 1.9 years). The experiment used a virtual reality setup in a normally illuminated room (Fig. 4.1). This study consisted of a main study and a control study, and both included three conditions. In the main study, all ten participants were fitted with a 3D head-mounted display and performed the task under three conditions: 0° (0° condition), 30° (30° condition), and 60° (60° condition). Each arm position (0°, 30°, and 60°) was maintained for 30 min, and thus the experiment was 1.5 h. Every 3 min, the participant was instructed to use the left arm to demonstrate (a) the perceived position of the right arm and (b) the position of the right arm on the display. The former was used to examine the effects of visual information on proprioception, and the latter was used to examine the effects of proprioception on visual information.

The experimenter moved the arm into 0°, 30°, or 60° of transverse adduction in a pseudo-randomized order. The participants' right forearm and upper arm were supported by the experimenter's hand as their arm was abducted or adducted. In all conditions, the right elbow joint was rigid and in full extension, so the arm was always straight, and the arm was parallel to the floor. A video camera (HDR-TD20V, Sony) was fixed above the right acromion at the height of the participant's eyes and used to record the right arm (see Fig. 4.1a). The orientation of the video camera was always pointed along the line of the arm. The video shot by the camera was presented on the display in real time. Regardless of the actual position of the participant's arm, the arm was presented on the display as extended in front of the trunk (i.e., 90° transverse adduction) (Fig. 4.1b). The discrepancy between visual and proprioceptive information on arm position therefore decreased from the 0° condition to the 60° condition (Figs. 4.2a–c). In the measurements of posture, the participants abducted the left arm to the height of the shoulder, and then both arms were put on the table (Fig. 4.1a). From this posture, the participants demonstrated the perceived posture of the right arm (reference arm) by a single movement of the left arm (indicator arm) when they performed for each of all three conditions. Five participants first indicated the perceived right shoulder position, and the remaining five first indicated the position of his right arm on the display.

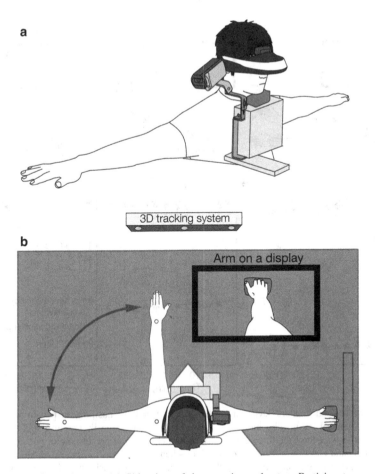

Fig. 4.1 Experimental setup. (**a**) Side view of the experimental setup. Participants were fitted with a 3D head-mounted display. A video camera was fixed above the right acromion to record a video of the right arm. The recorded video was presented on the display. To fix the participant's head and trunk, the chin was placed on a chin rest, and the trunk was attached to the rest. (**b**) *Top view* of the virtual reality setup and a seen arm position presented on the 3D head-mounted display. Participants were seated on a chair facing the 3D tracking system. For the measurements of posture, the participant abducted both their arms to the height of the shoulder and then put their arms on a table. Regardless of the position of the actual arm, the arm on the display appeared as though it extended in front of the trunk (i.e., 90° transverse adduction). To block visual information other than the arm on the display, a blackboard was placed in front of the arm. The participant's right hand was put on a black cushion to prevent the fingertips from obscuring the wrist (Masumoto and Inui 2015, Redrawn with permission from the Journal of Motor Behavior)

Participants were instructed to maintain attention on the video during the main study. The participants were looking at the display also in the moments they responded while they completed one trial (i.e., a single movement) in each condition. Participants were instructed not to shift from each stationary position

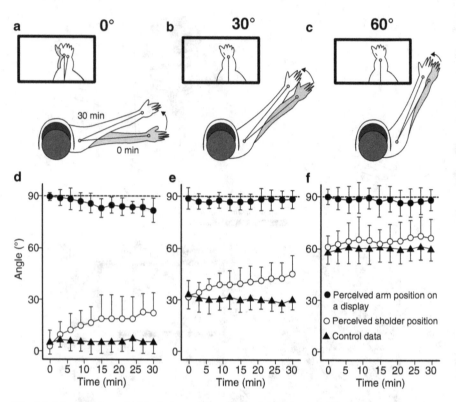

Fig. 4.2 Means and standard deviations of the perceived shoulder position, the perceived position of the seen arm on the display, and the perceived shoulder position in a control study over the 30-min period in each of the three conditions. (**a–c**) Schematic of perceived arm positions at the start (0 min) and end (30 min) of the experiment when the arm was transversely adducted to 0° (**a**), 30° (**b**), or 60° (**c**). The *gray arm* in the *square* indicates the position of the seen arm on the display, and the *gray arm* below the square indicates the true position of the arm. The *white arm* indicates the perceived position of the arm. (**d–f**) The perceived shoulder position (*open circles*), the perceived position of the seen arm on a display (*solid circles*), and the perceived shoulder position in the control study (*solid triangles*). *Symbols* indicate the mean and *error bars* indicate the standard deviation of ten participants. In the 0° condition (**d**), the perceived arm position (*open circles*) gradually approached the visually displayed arm position (the *gray arm* in the square) by 15 min and then stabilized at a value between the actual and fake positions. In the 30° condition (**e**), the perceived arm position (*open circles*) gradually approached the visually displayed arm position throughout the 30-min period. The perceived arm position more closely approached the visually displayed arm position as the difference between the actual and visually displayed arm positions increased. When the difference between the actual and visually displayed arms was 90° (**a, d**), the position of the seen arm on the display (*solid circles*) was perceived as approaching the position of the actual arm (Masumoto and Inui 2015, Redrawn with permission from the Journal of Motor Behavior)

during the 30-min period, and the experimenter confirmed every 3 min that the tip of the right index finger had not shifted from the target position. When the participants were not demonstrating the position of their right shoulder or arm with their left arm, the left arm was put on a black table with the left elbow joint flexed at 45° to prevent the arm position from affecting the demonstration of the right shoulder or arm position. Although the participants were unable to look at their actual arm, they were aware that their shoulder was fixed at different positions in each of three conditions.

A control study was performed on the same ten participants and revealed that there was no significant adaptation of perceived arm position over a 30-min period in any of the three test postures (see Figs. 4.2d–f). The participants in this control study were blindfolded. The experimenters moved their right arm into 0°, 30°, or 60° of transverse adduction in the same order as in the main study, and the participant used the left arm to demonstrate the perceived position of the right arm at the start of the control study and at 3-min intervals thereafter. Although a couple of participants showed a drift in perceived arm position toward the body midline during the 30-min period, mean data across all ten participants did not show any drift. This result is not consistent with that of Wann and Ibrahim (1992) but is consistent with that of Desmurget et al. (2000).

Participants were always seated on chairs. To fix their heads and trunks and enable accurate measurement of posture, the chins were placed on a chin rest, and the trunk was attached to a rest (Fig. 4.1a). In the main study, all participants were fitted with a 3D head-mounted display (HMZ-T1, Sony, Tokyo). The display was attached to the participant's face with a rubber light shield around the rim to obstruct light from the outside. To block all visual information except for the arm shown on the display, a blackboard was placed in front of the arm. The participant's right hand was placed on a black cushion to prevent his fingertips from being obscured from camera view by the wrist.

The position of the acromion and wrist was tracked by an infrared tracking system (OptiTrack V120, SenzTech, Beijing, China) at 120 Hz. The output of the system was recorded by a personal computer. Markers were attached on the left acromion and wrist in both the main and control studies. The angle between two markers was measured in the transverse plane (e.g., transverse abduction/adduction). This setup allowed us to compute the position of the acromion and wrist at sub-0.5-mm precision.

The perceived arm position more closely approached the visually displayed arm position as the difference between the actual and visually displayed arm positions increased. This resulted in absolute changes in perceived arm position but not changes relative to size of the distortion. If the difference between the actual and visually displayed arm positions was 90° throughout the 30-min period (i.e., the 0° condition; see Fig. 4.2a), the perceived arm position gradually approached the visually displayed arm position over the first 15 min and then stabilized at a value between the actual and visually displayed arm position. This indicates that proprioceptive estimate adapted to match the visual estimate. By contrast, when the difference between the actual and visually displayed arm positions was 90°, the visually displayed arm position was very gradually perceived as approaching the position of the actual arm as time passed (Fig. 4.2a). This indicates that visual estimate of the arm position adapted to match the proprioceptive estimate in the extreme case of a 90° discrepancy.

A novel result of this study is the slow change in perception over the course of the 30-min session. Whereas the proprioceptive estimate of arm position gradually adapted to match the visual estimate, the visual estimate was very gradually adapting to match the proprioceptive estimate. The visual estimate had not finish adapting to match the proprioceptive estimate at 30 min because there was a downward trend over time in the perceived position of the seen arm on the display (solid circles, Fig. 4.2d). Neither estimates gravitated toward some weighted position over a long period. Rather, the proprioceptive shift was larger than the visual shift. Presumably, the reason that the effect on proprioceptive adaption may be larger than the effect on visual adaptation is that the visual display is in the forward direction. If the virtual display showed the seen hand to be directed to the angles of 0°, 30°, and 60° while the actual hand was at 90°, the effect of visual adaption may be larger. Because the arms would not be usually visible in the 0° or 30° direction, we are used to seeing our hand in that forward workspace and may be more prone to estimate our arm to be in that direction. However, we were unable to direct the visually displayed hand to the angles of 0°, 30°, and 60° in the experimental setup of the present study.

Another novel result of this study is the adaptation to large rather than small discrepancies. Proprioceptive and visual adaptations to a discrepancy increased in absolute changes in perceived arm position with increasing magnitude of the discrepancy between these modalities. An important factor for the rubber hand illusion is the anatomical posture of the rubber hand. A rubber hand turned 180° (Ehrsson et al. 2004), 90° (Tsakiris and Haggard 2005), or even small deviations (Costantin and Haggard 2007) did not elicit the illusion. Ide (2013) recently tested the effect of posture on the rubber hand illusion. Participants perceived higher ownership of a rubber hand simulated by 3D computer graphics when the angle was at 0°, 45°, 90°, and 315° than at 180°, 225°, and 270°. These findings suggest that because body representation includes the knowledge of "anatomical plausibility" (Ide 2013), the brain eliminates a visually displayed limb that is presented in extreme postures from the body image. In contrast to these previous findings, the present study highlights that adaptation increases with increasing discrepancy between the actual and the visually displayed arm. One plausible contributory mechanism is that because the visually displayed arm was always presented on the head-mounted display as extended in front of the trunk and therefore did not involve postures that were anatomically impossible, the brain was able to adapt the large discrepancy to the body image. Moreover, whereas the rubber hand illusion is rapidly elicited from the integration of visual, tactile, and proprioceptive information, the discrepancy between visual and proprioceptive information in the present study was slowly resolved over the 30-min period even without visuotactile stimulation.

In future studies, we plan to confirm the effects of visual information of inverted hand images and arrow images (Ide and Hidaka 2013; Igarashi et al. 2007) on the resolution of a discrepancy of the hand position across sensory modalities using a virtual reality environment. In addition, we will examine the resolution of a discrepancy of the wrist position across sensory modalities using a virtual reality environment and then confirm the effects of ischemic nerve block of proprioception in the hand on the perceived wrist position (Inui et al. 2011; Inui and Masumoto 2013).

References

Costantin M, Haggard P (2007) The rubber hand illusion: sensitivity and reference frame for body ownership. Conscious Cogn 16:229–240

Desmurget M, Vindras P, Grea H, Viviani P, Grafton ST (2000) Proprioception does not quickly drift during visual occlusion. Exp Brain Res 134:363–377

Ehrsson HH, Spence C, Passingham RE (2004) That's my hand! Activity in premotor cortex reflects feeling of ownership of a limb. Science 305:875–877

Ghahramani Z, Wolpert DM, Jordan MI (1997) Computational models for sensorimotor integration. In: Morasso PG, Sanguineti VS (eds) Self-organization, computational maps and motor control. North-Holland, Amsterdam, pp 117–147

Harris CS (1963) Adaptation to displaced vision: visual, motor or proprioceptive change? Science 140:812–813

Ide M (2013) The effect of "anatomical plausibility" of hand angle on the rubber-hand illusion. Perception 42:103–111

Ide M, Hidaka S (2013) Visual presentation of hand image modulates visuo-tactile temporal order judgment. Exp Brain Res 228:43–50

Igarashi Y, Kitagawa N, Spencer C, Ichihara S (2007) Assessing the influence of schematic drawings of body parts on tactile discrimination performance using the crossmodal congruency task. Acta Psychol 124:190–208

Inui N, Masumoto J (2013) Effects of visual information on perceived posture of an experimental phantom foot. Exp Brain Res 226:487–494

Inui N, Walsh LD, Taylor JL, Gandevia SC (2011) Dynamic changes in the perceived posture of the hand during ischaemic anaesthesia of the arm. J Physiol 589:5775–5784

Martin TA, Keating JG, Goodkin HP, Bastian AJ, Thach WT (1996) Throwing while looking through prisms. II. Specificity and storage of multiple gaze-throw calibrations. Brain 119:1199–1211

Masumoto J, Inui N (2015) Visual and proprioceptive adaptation of arm position in a virtual environment. J Mot Behav 47:483–489

Redding GM, Wallace B (1996) Adaptive spatial alignment and strategic perceptual-motor control. J Exp Psychol Hum Percept Perform 22:379–394

Smeets JBJ, van den Dobbelsteen JJ, de Grave DDJ, van Beers RJ, Brenner E (2006) Sensory integration does not lead to sensory calibration. Proc Natl Acad Sci U S A 103:18781–18786

Tsakiris M, Haggard P (2005) The rubber hand illusion revisited: visuotactile integration and self-attribution. J Exp Psychol Hum Percept Perform 31:80–91

van Beers RJ, Sittig AC, Denier van der Gon JJ (1998) The precision of proprioceptive position sense. Exp Brain Res 122:367–377

van Beers RJ, Sittig AC, Denier van der Gon JJ (1999) Integration of proprioceptive and visual position-information: an experimentally supported model. J Neurophysiol 81:1355–1364

van Beers RJ, Wolpert DM, Haggard P (2002) When feeling is more important than seeing in sensorimotor adaptation. Curr Biol 12:834–837

Wann JP, Ibrahim SF (1992) Does limb proprioception drift? Exp Brain Res 91:162–166

Warren DH, Cleaves WT (1971) Visual-proprioceptive interaction under large amounts of conflict. J Exp Psychol 90:206–214

Welch RB, Warren DH (1986) Intersensory interactions. In: Boff KR, Kaufman L, Thomas JP (eds) Handbook of perception and human performance, vol 1. Wiley, New York, pp 25, 1–25, 36

Chapter 5
Conclusion

Abstract The brain needs body image to plan movements. A series of our studies uses an ischemic block of a limb to study the mechanisms of changes in body image. First, if the fingers, wrist, elbow, ankle, and knee are extended before and during the block, then the perceived limb is flexed at the joint and vice versa. The key parameter for these illusory changes in limb position may be the difference in discharge rates between afferents in flexor and extensor muscles at a joint. Then, the final position of the phantom limb depends on its initial position, suggesting that a body image refers incoming proprioceptive information for determination of starting points and end points during the generation of movements. In addition, the change in position does not involve limb postures which are anatomically impossible, suggesting that illusory posture is constrained by body maps or representations. Second, the perceived size of the hand increases gradually as anesthesia develops. The changes in both perceived hand size and perceived position of the joints develop when large-fiber cutaneous sensation is beginning to degrade. Hence, it is unlikely that a block of small-fiber afferents is critical for phantom formation in an ischemic block. Third, at the end of the block, when participants are allowed to see their foot, its perceived position reverts to that indicated by them earlier. The result indicates that visual information overrides the proprioceptive and tactile information in the formation of body image.

Keywords Body image • Phantom limb • Ischemic anesthesia • Proprioception • Vision

Main findings in a series of our studies are that if the fingers, wrist, elbow, ankle, and knee are extended fully before and during an ischemic block, then the body parts finally are perceived as flexed at the joints. Conversely, if the body parts are flexed fully, then the perceived position of the body parts moves toward extension (Inui et al. 2011, 2012b; Inui and Masumoto 2013). However, if the hand is held in the mid-position before and during the ischemic block, the posture of the wrist is perceived to be in the same position (Inui et al. 2012b). Then, after participants look at their foot and leg blocked by cuff inflation at the end of the ischemic block, the perceived posture of both the ankle and knee returns to the position perceived at 0–25 min after the block (Inui and Masumoto 2013). In addition,

© The Author(s) 2016
N. Inui, *Systematic Changes in Body Image Following Formation of Phantom Limbs*,
SpringerBriefs in Biology, DOI 10.1007/978-981-10-1460-4_5

in a visually displayed arm position, whereas the proprioceptive estimate of arm position gradually adapted to match the visual estimate, the visual estimate was very gradually adapting to match the proprioceptive estimate (Masumoto and Inui 2015). On the other hand, while the right hand and forearm constituted an incomplete phantom limb from 10 to 20 min after the ischemic block, the stick fixed to the hand was perceived as having moved. Such a new type of hand–object illusion was found on the base of the rapid changes of the cortical circuits (Inui and Masumoto 2015).

The brain holds an image of the limb which depends on the background sensory input. As input from somatosensory receptors fades during ischemic anesthesia, the perceived limb moves away from its initial position. At an extreme posture, one signal will be high and the other very low so that disappearance of afferents will mainly alter the high signal. As the anesthesia progresses, the stretch receptor activity falls, and this fall is interpreted by participants as a change in limb posture. In contrast, in mid-position the firing of afferents signaling flexion and extension is balanced so that as afferent firing disappears it does so in a balanced way. These observations highlight that the body image as an online body representation is updating by the moment-to-moment input coming from the body periphery. The key parameter for the illusory changes in limb position may be the difference in discharge rates between afferents in flexor and extensor muscles at a joint.

A series of our studies (Inui et al. 2011, 2012a, b; Inui and Masumoto 2013) indicates that the changes in posture of the limb depend on one posture immediately before inflation. This finding suggests that a body image refers incoming proprioceptive information for determination of starting points and end points during the generation of movements. Then, our studies consistently indicate that the change in position does not involve limb postures which are anatomically impossible. This observation suggests that illusory posture is constrained by body maps or representations. In addition, although the limb remained completely stationary during the period of cuff inflation in our study, its perceived posture changed progressively. After 30–40 min, passive movements of the limb go undetected, yet, if participants command their wrist to flex or extend, the phantom shifts its position in the direction of the intended movement (Gandevia et al. 2006; see also Smith et al. 2009), and the speed of this illusory movement grades with the level of subjective effort (Walsh et al. 2010). This highlights the ease with which the "phantom" hand is incorporated into the systems which plan and execute movement.

Goodwin et al. (1972) find that if vibration is applied over the belly and the tendon of biceps while the arm is being moved by the experimenter, participants consistently overestimate the angle of extension at the elbow. Albert et al. (2006) and Roll et al. (2009) further record the actual proprioceptive feedback from a series of muscles while participants performed movements. Then the firing frequencies of the muscle spindle afferents are translated into a vibration frequency, and the vibration is applied to participants. The illusory movements are mimicking writing or drawing movements. These studies on vibration-induced illusions of movements indicate that awareness of movements matches quite well the information that is present in the muscle spindles. Neuroimaging studies (Naito et al. 1999, 2002) show that the primary motor cortex codes the sensation of movements in the presence of vibration-induced illusions of movements in spite of no performance of movements. In our studies, because a body image constantly refers incoming proprioceptive information, the results are in part similar to those of the studies on vibration-induced illusions of movements. In contrast to our studies, Craske (1977) shows vibration-induced illusions of movements that are biomechanically impossible to achieve. In addition, Lackner (1988) asks participants to pinch their nose with their index finger and thumb and vibrate their biceps tendon. The participants then feel that their nose becomes longer (also see Naito et al. 2002), indicating that the Pinocchio illusion involves a body shape which is anatomically distorted.

On the other hand, Longo and Haggard (2010) point out that although information about body posture is specified by ongoing afferent inputs, no sensory signal is available to detail body size and shape. And they suggest an additional map incorporating the body's metric properties. Although the map is distorted and crudely resembles the maps drawn by Penfield and Boldrey (1937), participants consciously perceive only the undistorted map. Walsh et al. (2015) recently show that the changes in perceived body size following local anesthesia are not uniform. They found that during anesthesia the finger is perceived to get up to 30 % fatter, but less than 3 % longer. It is expected that the different body forms and the distortions are examined experimentally in the future.

References

Albert F, Bergenheim M, Riot-Ciscar E, Roll J–P (2006) The Ia afferent feedback of a given movement evoked the illusion of the same movement when returned to the subject via muscle tendon vibration. Exp Brain Res 172:163–174

Craske B (1977) Perception of impossible limb positions induced by tendon vibration. Science 196:71–73

Gandevia SC, Smith JL, Crawford M, Proske U, Taylor JL (2006) Motor commands contribute to human position sense. J Physiol 571:703–710

Goodwin GM, McCloskey DI, Matthews PB (1972) The contribution of muscle afferents to kinaesthesia shown by vibration induced illusions of movement and by the effects of paralysing joint afferents. Brain 95:705–748

Inui N, Masumoto J (2013) Effects of visual information on perceived posture of an experimental phantom foot. Exp Brain Res 226:487–494

Inui N, Masumoto J (2015) Perceptual changes of interaction between hand and object in an experimental phantom hand. J Mot Behav 47:81–88

Inui N, Walsh LD, Taylor JL, Gandevia SC (2011) Dynamic changes in the perceived posture of the hand during ischaemic anaesthesia of the arm. J Physiol 589:5775–5784

Inui N, Masumoto J, Beppu T, Shiokawa Y, Akitsu H (2012a) Loss of large-diameter nerve sensory input changes the perceived posture. Exp Brain Res 221:369–375

Inui N, Masumoto J, Ueda Y, Ide K (2012b) Systematic changes in the perceived posture of the wrist and elbow during formation of a phantom hand and arm. Exp Brain Res 218:487–494

Lackner JR (1988) Some proprioceptive influences on the perceptual representation of body shape and orientation. Brain 111:281–297

Longo MR, Haggard P (2010) An implicit body representation underlying human position sense. Proc Natl Acad Sci U S A 107:11727–11732

Masumoto J, Inui N (2015) Visual and proprioceptive adaptation of arm position in a virtual environment. J Mot Behav 47:483–489

Naito E, Ehrsson HH, Geyer S, Zilles K, Roland PE (1999) Illusory arm movements activate cortical motor areas: a positron emission tomography study. J Neurosci 19:6134–6144

Naito E, Roland PE, Ehrsson HH (2002) I feel my hand moving: a new role of the primary motor cortex in somatic perception of limb movement. Neuron 36:979–988

Penfield W, Boldrey E (1937) Somatic motor and sensory representation in the cerebral cortex of man as studied by electrical stimulation. Brain 60:389–443

Roll J–P, Albert F, Thyrion C, Riot–Ciscar E, Bergenheim M, Mattei B (2009) Inducing any virtual two-dimensional movement in humans by applying muscle tendon vibration. J Neurophysiol 101:816–823

Smith JL, Crawford M, Proske U, Taylor JL, Gandevia SC (2009) Signals of motor command bias joint position sense in the presence of feedback from proprioceptors. J Appl Physiol 106:950–958

Walsh LD, Gandevia SC, Taylor JL (2010) Illusory movements of a phantom hand grade with the duration and magnitude of motor commands. J Physiol 588:1269–1280

Walsh LD, Hoad D, Rothwell JC, Gandevia SC, Haggard P (2015) Anaesthesia changes perceived finger width but not finger length. Exp Brain Res 233:1761–1771

Acknowledgments

This monograph was supported by the Japan Society for the Promotion of Science (21500544 and 24500679) and the National Health and Medical Research Council of Australia.

Printed in the United States
By Bookmasters